What's in This Book . . .

Can foods heal? Does what we eat really affect the way we feel and function? Inside, the writings of three teachers and healers—Hippocrates. V. G. Rocine, and Bernard Jensen—offer compelling evidence that what we put in our mouths has a profound effect on our health and well-being.

Based upon three lifetimes of observation, study, and research, the fascinating principles presented by these men in Part One may change the way you look at your next meal. In addition to these basic concepts, the first part of the book contains a host of helpful troubleshooting advice: health cocktails for common ailments, herbal teas, tonics, vitamin- and mineral-packed food combinations, and detailed data on the roles foods play in the optimum efficiency of specific bodily systems, functions, and overall health.

Part Two of "Foods That Heal" provides an easy-to-understand guide to fruits and vegetables. Each listing in this section presents a history of use, a buyers' guide, therapeutic benefits, and nutrient information.

Both those looking to improve their health and those interested in taking an active role in enhancing their overall well-being will find this book interesting, informative, and full of common-sense suggestions for attaining—and maintaining—good health through proper nutrition.

Dr. Bernard Jensen

A GUIDE TO UNDERSTANDING AND USING
THE HEALING POWERS OF NATURAL FOODS

FOODS
THAT HEAL

Dr. Bernard Jensen

AVERY PUBLISHING GROUP INC.
Garden City Park, New York

The medical and health procedures in this book are based on the training, personal experience, and research of the author. Because each person and situation is unique, the editor and the publisher urge the reader to check with a qualified health professional before using any procedure where there is any question as to its appropriateness. The publisher does not advocate the use of any particular diet, but believes the information presented in this book should be available to the public.

Because there is always some risk involved, the author and publisher are not responsible for any adverse effects or consequences from the use of any of the suggestions, preparations or procedures in this book. Please do not use the book if you are unwilling to assume the risk. Feel free to consult a physician or other qualified health professional. It is a sign of wisdom, not cowardice, to seek a second or third opinion.

Cover Design: Martin Hochberg and Rudy Shur
Cover Photo: Ultimate Image, Inc.
In-House Editors: Karen Price Heffernan,
                 Nancy Marks Papritz
Typesetting: Multifacit Graphics, Inc.

Library of Congress Cataloging-in-Publication Data

Jensen, Bernard, 1908-
    Foods that heal.

    Includes index.
    1. Nutrition. 2. Health. 3. Fruit. 4. Vegetables.
I. Title.
R.A784.J46   1988   613.2   88-7383
ISBN 0-89529-405-2

Printed in the United States of America

10 9 8 7 6 5 4

Contents

To my dearest friends,
Lynne, Eleanor, and their son Larry,
all healers and all a credit to the healing arts.

Preface

I am not the kind of physician who performs surgery, prescribes or administers drugs, or practices medicine the way most modern physicians are generally thought to do. Rather, for the past fifty-five years, I have been a different kind of physician. I have counseled patients, striving to guide and uplift them by building their health and teaching them that there is a right way and a wrong way to live.

This does not mean I have not taken care of sick people. Hundreds of thousands of patients have entered my sanitariums—many with serious chronic disease—some in wheelchairs, several on stretchers. I have had the privilege of seeing the great majority of them leave free of the symptoms and conditions that brought them into my care.

I have treated these patients using a combination of proper nutrition, exercise, positive thinking exercises, water treatment, and other natural methods. Though I believe in the scientific merit of certain therapeutic drugs, I do not advise their use nor do I use them myself. Though I believe surgery has its place in the treatment of certain life-threatening diseases and extreme conditions, I advocate the use of less invasive, more natural methods in most cases.

I do not regard the healing art lightly. On the contrary, taking care of people has been both my life's ideal and its privilege. I sincerely feel that each person I treat is a living soul and a member of the family of man and, as such, is entitled to love and respect. As a physician, I feel a humanitarian responsibility to respond to suffering and its needs.

The story of how I developed my philosophy begins in 1926 when I was a young man of 18. It was then that I entered the West Coast Chiropractic College, supporting myself by working at a local dairy. Long hours of study, followed by long hours of work, combined with poor nutritional habits, posed a triple threat to my health. Shortly after my graduation from college I collapsed.

Physicians diagnosed my condition as bronchiectasis, an incurable lung disease, often fatal in those days before antibiotic treatment. I had inherited weak lungs from my mother, who died of tuberculosis at the age of 29. Lung weakness ran in my family, and now it had run into me.

It was about this time that I was introduced to a Seventh-Day Adventist physician who enlightened me on the differences between a poor food regimen and a healthy one. Sadly, his name escaped me over the years. I certainly owe him a debt of gratitude because of the path he set me upon. This doctor declared that a root of my problem was my nutritional deficiencies. I was, he said, starving myself with a "junk food" diet. In its place he prescribed a diet full of healthy foods. Combined with breathing exercises given by Thomas Gaines who once worked for the New York Police Department, my condition improved. I began to gain weight, put several inches of flesh back on my chest, and found renewed energy. I was back on the road to health.

Though I began my career in the health arts as a chiropractor, my remarkable experience with the regenerative abilities of proper nutrition and exercise spurred me to incorporate these healing methods in my growing practice. In addition, I continued my postgraduate education to keep abreast of new developments in natural health care. I worked along side Dr. Ralph Benner of the Bircher-Benner Clinic in Zurich, Switzerland. I studied bowel management with Dr. John Harvey Kellogg of Battle Creek, Michigan; iridology with Dr. R. M. McClain of Oakland, California and Dr. F. W. Collins of Orange, New Jersey; homeopathy with Dr. Charles Gesser of Tampa, Florida; and water cure treatment at Bad Wörishofen, West Germany, home of nineteenth-century water therapy pioneer Fr. Sebastian Kneipp.

Now, at the age of 80, I often reflect on what it was in my life that allowed me to live this long—to come this far. For though I had cured bronchiectasis with nutrition and exercise, I continued the frantic pace of work and study that, combined with my bad habits, had made me so ill so long ago. Looking back, I have concluded that wellness is as much a satisfying relationship with life as it is a consequence of dietary and lifestyle changes.

I believe the secret of my good health is that I am always good to myself mentally. I am convinced my longevity is due to my mental philosophy, my joyous contentment with life. I have always loved people. I have always loved seeing people who came to me for help return home healed. And these people loved me in return. I uplifted them to the best of my ability, and it always came back to me.

Perhaps it was the warmth of this kind of gratitude that provided the incentive and energy for me to do so much more than I had to do. When the people you take care of want to take care of you, life becomes a blessing. I feel I had blessings that uplifted me constantly.

For this reason I have come to believe that loving your work is one of the great secrets of health and high-level well-being. On most of my twelve-to-fourteen-hour-long days at sanitariums, including my ranch in California, I have never felt overworked or "burned out" at the end of the day. And each morning I awake eager to get going again because I love my work.

I have received many honors and awards during my lifetime, but the greatest gift I have ever been given is the gratitude, love, and respect of the thousands of patients whose lives have been changed by what I have been able to share with them. It is through serving them that I have found the greatest portion of my own life's happiness.

Introduction

F rom the time of Hippocrates, it has been known that certain foods have disease-preventing and disease-healing benefits. Yet, for some reason this knowledge has remained one of history's best-kept secrets—until its resurgence in very recent years.

Despite the growing body of documented medical evidence that diet both causes and cures disease, nutritional awareness remains far from a twentieth century world ideal. According to Stuart M. Berger, M.D., in *What Your Doctor Didn't Learn in Medical School*, currently a mere 24 of this country's 130 medical schools require future doctors to take courses in nutrition. By omitting the subject of nutrition, 80 percent of America's medical schools are not only perpetuating a nutritional "knowledge vacuum," they are sending out a negative message about the importance of nutrition in health, as well.

With our doctors ill-educated on nutrition, it's no wonder the public continues to lag in its own nutritional awareness. For instance, a recent survey of 12,000 Americans by the National Cancer Institute (NCI) revealed a majority continues to practice poor eating habits, despite strong evidence that diet can reverse the course of some forms of cancer, heart disease, diabetes, and learning disorders—to name a few. Asked by NCI what they had eaten in the last 24 hours, more than 40 percent of the respondents said they had not had even one piece of fruit, and about 20 percent said they had not eaten even one vegetable. Just as discouraging, some 55 percent of the NCI survey group said they had eaten red meat, and more than 40 percent had at least one serving of luncheon meat or bacon that day.

Results like these lead me to believe that the message of good nutrition is not getting out there. Because of this, it is my hope that this book becomes a nutritional guidebook for the common man's journey toward disease-free living through proper nutrition.

Far from a "bandage" approach like medicine applied to a wound, proper nutrition changes the course of disease at its source: tissue structure. No therapy or drug known to modern medical science can rebuild tissue that has been damaged by disease or trauma. Food alone can accomplish this feat. It is for this reason that nutrition is an indispensable weapon against disease.

But the story of nutrition is not simply one of cure. It is also a story of life-enrichment and well-being. Sadly, many people are living at only 50 percent of their full health potential, not really sick, but not truly well either. These people need to understand that the same foods that heal by rebuilding damaged tissue will enhance wellness by increasing the efficiency and energy level of underactive endocrine glands, and all other organs, glands, and tissues—including the skin, the muscles, the nerves, the joints, the veins, and the arteries. The message is clear: You can feel wonderful—if you will simply eat healthful foods and avoid harmful foods.

For 55 of my 80 years, my life has centered on the application of foods to healing and wellness. In this book, I discuss all I've learned about nutrition, healing, and wellness so that you will have an opportunity to enjoy your own life and health as I am enjoying mine.

Chapter One of this book is dedicated to a discussion of the great physician Hippocrates' influence on modern health care from a nutritional standpoint. Before Hippocrates' lifetime during Greece's Golden Age, health care was still a hodgepodge of superstitious rituals. Indeed, after his time, much of the medical knowledge he gave Western civilization sank back into obscurity for centuries. (Some claim the nutritional knowledge he pioneered remains there still.) Hippocrates was probably the first physician to employ observation, analysis, and practical procedures such as diet change to promote healing in his patients. In addition, this greatest of physicians was committed to ethics in the physician-patient relationship. I use many of Hippocrates' practices and principles in my own work.

Though less well-known than Hippocrates, Dr. Victor G. Rocine was equally influential in my work. Rocine emigrated from

Norway to the United States after studying the work of pioneer European food chemists—work that had not yet been introduced to this country. To Rocine, all illness and disease could be traced to either nutrient deficiencies or excesses in the human body. Rocine "rediscovered" the preventive approach to disease first advocated by Hippocrates more than 2,000 years before. For this reason, I have dedicated Chapter Two to the work of Rocine.

Much of Chapters One and Two comes from the original works of Hippocrates and Dr. Rocine.

The impact of these two great teachers on my own work is discussed in detail in Chapter Three, a discussion of my work in the health arts. A brief concluding chapter suggests how to get started making changes.

Following Part One's narrative chapters, which deal with the basis of my work, I have included Part Two, entitled *A Guide to Fruits and Vegetables*. This section will provide you with a handy home reference guide to the actions and importance of scores of foods. I am sure you will find it a valuable tool to gain better understanding about the way foods can help us and heal us. Combined with the knowledge in Part One, it is my hope that this book as a whole will serve as a nutritional source book to guide you and your family towards better health.

Part One

Three Pathways
to Health

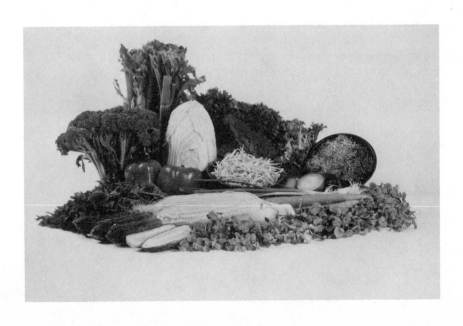

"Let food be thy medicine."

—*Hippocrates*
circa 431 B.C.

1.

Hippocrates and His Work

I t is because Hippocrates laid the cornerstone for modern medical and nutritional science that I discuss his work here first. I have been fascinated with the teaching of Hippocrates for as long as I can remember. The reason is not only that he believed in foods and natural cure, but also that he was committed to serving his fellow man, and that he applied wisdom and integrity to his art. In other words, I like his philosophy as well as his methods.

Hippocrates believed in the power of positive thinking. "Some patients, though conscious that their condition is perilous, recover their health simply through their contentment with the goodness of the physician," he wrote.

Over the years I have found, as Hippocrates taught, that my relationship with patients is just as important as the attention I give to their health. This is because when people trust you, they seem to get well more consistently. But the physician's concern must be genuine, Hippocrates wisely observed. For the patient will only trust the physician fully when he or she feels the physician is giving his very best.

Hippocrates taught that the physician must be full of personal integrity. "Sometimes give your services for nothing," Hippocrates advised his students. "Unless the doctor values doing what is right more than he values money, the ideal of service to humanity is corrupted and the high purpose of the health art is compromised."

Called the Father of Medicine by historians, Hippocrates is

3

believed to have been born on the Greek island of Cos around 460 B.C. One biographer suggests he was practicing medicine by 431 B.C., at the age of 29. Historians believe that Hippocrates continued practicing his art well into his eighties, to about 377 B.C., traveling to various parts of the Greek world.

The real greatness of Hippocrates was not only in turning the healing arts away from magic and superstition to a more scientific approach, but in recognizing that therapy must be consistent with nature and the design of the human body. He knew that effective health care could not be separated from nutrition as part of the therapy. Because he has been widely respected through the ages, some of his writings, such as the *Aphorisms*, were used in medical schools throughout the world until as recently as the 1800s. Curiously, however, Hippocrates' writings on foods have been all but ignored by the American medical mainstream.

OBSERVATION AND REFLECTION

The main instruments of value available to Hippocrates were his own eyes and a thoughtful mind. No stethoscope or thermometer was around in his time. Yet his comments on disease and symptoms are still considered advanced wisdom in the health arts.

"Every disease has its own nature and arises from external causes," Hippocrates wrote. He believed in the importance of methods of natural cure, such as diet. "Natural forces within us are the true healers of disease . . . Healing is a matter of time, but it is sometimes also a matter of opportunity . . . Everything in excess is opposed to nature . . . To do nothing is sometimes a good remedy." We are often surprised at how closely his philosophy parallels that of the modern wholistic health perspective.

"Unprovoked fatigue signals the presence of disease," he observed. "Overeating causes sickness, as the cure shows," he noted. One of his most well-known aphorisms is "Life is short, (the health) art is long, opportunity is elusive, experiment is dangerous, judgment is difficult. It is not enough for the physician to do what is necessary, but the patient and attendant must do their part as well, and circumstances must be favorable."

Hippocrates stressed prevention of disease by strongly recommending not only a balanced diet but a moderate and sensible lifestyle, as well. His essay on "Regimen" is the first known work on the preservation of health by a proper way of life—in other words, it is the first work on the value of preventive medicine. In addition, there were other unusually accurate observations. Hippocrates described air as entering the lungs, then the blood vessels, pioneering the concept of oxygenation of the blood.

Conversely, on some things Hippocrates was a bit off the mark. In his time, the identification and classification of what we call disease was not very far along. As a result, some diseases were named, some were not. Hippocrates was mainly interested in acute diseases, although he also recognized chronic stages of disease. For the most part, however, he limits his disease descriptions to chest conditions and fevers.

A CASE FOR MEDICINE

We find that Hippocrates possessed considerable insight into human nature. This is evident in his argument for medicine. "Sheer necessity has caused men to seek and to find medicine because sick men did not and do not profit by the same regimen as do men in health," he taught. There are some who refuse to follow standards of right living, so they lose their health, while others with weak constitutions and inherent weaknesses have to follow more intricate health guidelines than the average healthy person, simply to avoid disease, he indicated.

THE HIPPOCRATIC OATH

Hippocrates is perhaps most famous for his authorship of the *Hippocratic Oath*, adopted by the medical profession as a standard of ethical conduct. The most significant part of the oath, following its introduction, is:

> I will follow that method of treatment, which, according to my ability and judgment, I consider for the benefit of my patients, and abstain from what is deleterious and mischievous. I will give no deadly medicine to anyone if asked, nor suggest such counsel; furthermore, I will not give to a woman an instrument to produce an abortion. With purity and with holiness I will pass my life and practice my art. I will not cut a person who is suffering with a stone, but will leave this to be done by practitioners of this work. Into whatever houses I enter I will go into them for the benefit of the sick and will abstain from every voluntary act of mischief and corruption, and further, from the seduction of

females or males, bond or free. Whatever in connection
with my professional practice, or not in connection with my
professional practice, I may see or hear in the lives of men
which ought not to be spoken abroad I will not divulge, as
reckoning that all such should be kept secret. While I continue to keep this oath inviolate, may it be granted to me to
enjoy life and the practice of my art, respected always by all
men, but should I trespass and violate this oath, the reverse
be my lot.

HIPPOCRATES—AN UPDATED OVERVIEW

Following, in an easy-to-read format, are paraphrased and updated
selected portions of the works of Hippocrates to illustrate his philosophy and viewpoints in the health arts. As you read through
these sections, it may be helpful to view them in their historical
context: the Golden Age of Greece, circa 431 B.C.

INDIVIDUAL NEEDS VARY

Some people easily deal with strong foods. To others, strong foods
bring pain. The discovery that some foods cause harm while others
are beneficial can be considered the art of medicine, since it
emerged from a desire to promote the health, well-being, and nourishment of man. Those who study athletics and physical development are constantly making new discoveries concerning which
foods or drinks are best to strengthen a person.

No one would have attempted to discover medicine if the same
foods and lifestyle had been used both by the sick and the well. But
the fact is, any careful observer will find a difference.

The first medical treatment was probably a reduction in the
amount of food and drink allowed, without a change in diet. But,
although this treatment was beneficial to some patients, it was not
helpful in all cases, because some were in such poor condition that
they could not take even small amounts of food. Such patients were
thought to need weaker nutrients. "Soups" were formulated by
diluting full-strength foods with water and boiling them together.
Still others could only take clear liquid and no solid nutriment.

The amounts of food taken have as much of an effect on health
as the number of meals one takes. Some persons get along best with
one meal a day, others require two. Some, who take lunch when

food at this time does not suit them, become heavy and sluggish in body and mind, yawning and becoming drowsy and thirsty. If a healthy man takes insufficient food, there is just as much harm done as if he takes excess. Nutritional deficiencies are the cause of many evils, different from the problems caused by chronic gluttony, but just as harmful to the body.

COMMON DIETARY MISTAKES

One common mistake in food customs is to eat too much too frequently, taking the next meal before the previous one has been assimilated. This overworks the digestive organs. Others space their meals too far apart, depleting their energy stores, and bringing about fatigue before taking the next meal. This, too, places a strain upon the body and the health.

Persons with susceptible constitutions are more likely to experience ill effects from the preceding dietary mistakes than others, weakening more quickly than most other persons. And a weak person is only a step away from being a sick person, although a sick person is weaker still.

If the cook prepares every food the same way, there is no pleasure in the food. Nor would there be pleasure if he mixed all the foods together and served them in one dish.

The physician should be able to properly judge the differing effects of various foods and drinks. Some are astringent or laxative, others diuretic. Many foods are binding, drying, or moistening.

Barley is naturally cold, moist, and drying but contains a purgative in the juice of the husks. When unwinnowed barley is boiled and used, it is purgative. If winnowed barley is used, it is more cooling and astringent. When parched, the moist and purgative quality is eliminated by heat, and the remainder is cool and dry. Barley meal used with the bran passes better by stool.

A decoction of boiled bran is light and passes well by stool. Meal boiled in milk passes better by stool because of the whey, especially if mixed with laxatives.

Turnips are heating, moistening, and disturbing to the body. Pennyroyal and marjoram warm, as does savory, while thyme is hot. Hyssop is warming and expels phlegmatic humors. Diuretics include the juices of celery, garlic, clover, fennel, leeks, and maiden-hair. Cooling foods include mint, endive, bar-parsley, hypericum, and nettles. Juices that stimulate the bowel or purging are the juices of the chick-peas, lentils, barley, beet, cabbage, or elder.

Most physicians, like laymen, are likely to consider anything unusual done by a patient on the day of outbreak of an illness to be the cause. Extreme care should be used when drawing such conclusions. If the patient has taken a bath or a walk or has eaten some strange food, one of these may be mistakenly considered as the cause. Ignorant of the real cause, the physician must guard against assigning some treatment that is completely inappropriate.

OBSERVATIONS ON DISEASES

There are two kinds of diseases. Endemic diseases are always to be found, but epidemic diseases may come through the change in the seasons. Men with epilepsy are likely to live longer than others.

Women may become barren through the waters being hard, unassimilated, and cold. Their menstrual discharges become scanty and foul. Childbirth is difficult, although spontaneous abortion is seldom found. After birth of the child, the mother's milk dries up and she often develops tuberculosis.

PREDICTION AND DESCRIPTION OF DISEASE

I believe it is worthwhile for a doctor to be able to predict the course of a disease. A doctor who can declare to his patient the present state, past symptoms, and what is yet to come in a disease will have the full trust of his patient for treatment.

Now, to restore every patient to health is impossible. Men do die—some because of the severity of the disease before the doctor is called, others dying before the doctor can apply his art. It is necessary, therefore, to learn the nature of the diseases, how they may exceed the strength of men's bodies, and how to predict them. By doing so, you will win respect and high regard as a physician.

In acute diseases (those that are brief but severe), the doctor must examine his patients carefully. First, examine the face, then inquire whether the patient has been able to sleep, whether his bowel movements have been loose, and whether he still has an appetite. If any of these are present, the danger is less. In all cases, a healing crisis signifies the end of the disease.

Concerning chest conditions, good respiration must be considered to greatly improve the chance of recovery in acute diseases with fever in which a crisis is reached in forty days.

Heavy perspiration on critical days can completely get rid of fever in an acute condition. It is best when the whole body perspires. The worst sign is when perspiration breaks out only around the head and neck.

The patient ought to remain awake during the day and sleep only at night. It is bad if this order is not observed. The worst case is inability to sleep either day or night.

Bowel eliminations should be soft and uniform in texture, coming at the usual time of day as in health. The patient should eliminate from the bowel two or three times a day, depending on the amount of food taken, and once at night. The largest stool should be in the morning.

The best sign regarding the urine is when the sediment is white, smooth, and even throughout the illness to the time of crisis. This indicates a short sickness and a certain recovery.

Hardness and pain in the bladder are always bad, but they are worse when accompanied by constant fever. In such cases, stools will often be hard and constipation may be present.

Fevers come to a crisis on the same days, both those from which patients recover and those from which they die. The mildest fevers cease on the fourth day, or earlier. In all cases where fevers stop without sign of recovery or crisis, a relapse is indicated. When a disease shows signs of irregularity such as with a relapse, it is likely to be a long one.

CARING FOR ACUTE DISEASES

Always try to get the best doctor available for the most dangerous diseases. Many laymen are prone to judge as excellence an incompetent doctor's strange remedies.

There are many treatments available. Be cautious in your choice. For example, some doctors treating acute diseases prescribe unstrained barley gruel, while others consider that only the juice strained from the gruel may be taken, and still others withhold all food until the disease reaches a crisis.

Concerning nourishment, I think barley gruel is better than all other cereal foods in taking care of acute diseases. The finest barley should be used.

Concerning fasting, I am persuaded that doctors who fast their patients for two, three, or more days at the outset of a disease are doing the opposite of what they should. Perhaps they think it is natural to counteract one violent change in one's body—the disease—with another violent change—starvation. I do not agree.

When a patient, contrary to his usual routine, fasts for a day, he should avoid heat, cold, and fatigue, and he should break the fast with a small amount of moist food and a proportionate drink to go with it.

A physician must seriously study what is beneficial in a patient's diet and lifestyle while he is in good health. It is easily verified that a simple diet of food and drink, if persistently followed, is safer for health than a sudden violent change.

After fevers, those who get joint pains or tumors are taking too much food.

A proper regimen may include fermentations, baths, enemas, suppositories, and compresses. Some use is made of water and such drinks as hydromel (honey and water), oxymel (honey and vinegar), and wine, but the most common liquid used in acute diseases is barley gruel. Sometimes the pure strained juice is used, sometimes the solid barley is used, but nothing else is given until after the healing crisis.

The chief causes of diseases are violent changes in our usual life regimen and habits.

The physician should always consider the strength and character of each illness, the constitution of the patient, and the customary regimen of the patient—his food and drink.

Many doctors lack experience in distinguishing the various causes of weakness that occur in diseases—due to starvation, inflammation, pain, or the acuteness of the disease. The knowledge of such things may mean the difference between life and death.

It is one of the most serious blunders when a patient is weak through pain or the acuteness of the disease, to give food and drink under the impression that the weakness is from lack of nourishment. It is equally a shame when the physician fails to recognize weakness due to lack of nourishment. The latter is dangerous, although not as dangerous as the former.

Either resting too long or too much exertion while under treatment for a disease is injurious to the patient.

After long fasting, it is necessary to break the fast with a modest amount of food. If more than a modest amount is taken, there may be harm to the bowels.

Sleeplessness interferes with digestion, while too much sleep predisposes the body to flabbiness and the mind to stupefaction.

The drink oxymel (honey and vinegar) is often effective in acute diseases. It eases the breathing and brings up phlegm.

Water given in acute conditions neither soothes a cough nor brings up phlegm, as well as other things, in cases of pneumonia.

The bath will be good for many patients, long baths in some cases, short baths taken at regular intervals in other cases.

SEASONAL ADAPTATIONS

In developing a healthy lifestyle and diet, consider age, season, habits, land, and physique. Walking should be fast during winter and slow during summer. Bathe more often in summer, less in winter. The slender person should bathe more often than the heavier person. Overweight persons should eat their meals after exercise while they are still panting, before they have cooled off. They should have only one meal a day, refrain from bathing, sleep on a hard bed, and dress lightly. Thin people who want to gain weight should do the opposite of these things.

In no case disturb a patient while he is going through a crisis or just after one. Don't try to introduce purges or enemas, but leave the patient alone.

Nutrition can be harmful to a toxin-laden body.

When a patient has a strong appetite but his condition has not improved, that can be an indication of a turn for the worse.

People with healthy bodies lose strength when they take purges, as do those on a bad diet.

Excess and suddenness in bowel elimination, in taking foods, in warming or cooling the body, or in any other way disturbing it, is dangerous. "Little by little" is a good rule of thumb.

Left to themselves, patients lose hope during their painful suffering, give up the fight, and no longer resist death.

PRACTICING THE HEALTH ART

The physician must have at hand a sense of wit and humor, for a sour disposition is unwelcome among both the healthy and the sick. He must avoid gossip, fuss, or showiness, to avoid criticism.

A physician is justified if, in difficulty with a patient and in lack of experience, he urges the calling in of others to learn by consultation the truth about the case.

It is necessary to develop skill in palpation, anointing, washing, gracefulness in using the hands to place lint, compresses, bandages, ventilation, purges, and other things. Have ready beforehand the instruments, appliances, knives, and so on.

Make frequent visits. Be careful in your examinations, being prepared to deal with changes in the condition of the disease.

Keep a watch on the faults of patients who may try to avoid taking medicines repugnant to the taste, to their own harm.

Perform all tasks calmly and efficiently. Give orders calmly and cheerfully, reprove if necessary, and comfort the patient, if appropriate.

Physicians consulting with one another on a case must never quarrel or jeer at each other. I believe it is very important that a physician should never be jealous of another.

THE PHYSICIAN'S APPEARANCE

The dignity of a doctor requires that he should look healthy, or people will think he is unfit to care for patients. He must be clean, well-dressed, and anointed with a scent that is pleasing to patients.

A doctor must be a gentleman, grave in manner, and kind toward all. In appearance he should be serious but not harsh, neither should he be a man of uncontrolled gaiety, but avoiding vulgarity. Patients greatly trust their doctors, so toward women and maidens—and toward all—self-control must be used.

HIPPOCRATES IN PERSPECTIVE

When we recall that Hippocrates lived at the beginning of an era preceded by a good deal of reliance on superstition and magic, it is a wonderful thing to see how the ancient physician was able to discover the healing and life-changing properties of foods. For his time, Hippocrates' wisdom created a beacon of reason that shone over the centuries that followed.

Hippocrates is not so much known for his originality, as he is known for his wisdom in making sense of much of the basic knowledge of his time, and in putting it all together.

In my view, his knowledge of the healing crisis—that point in the disease where symptoms recur along the pathway to cure—makes him an advanced and perceptive doctor, not only for his own time but for ours as well.

Hippocrates was aware that many things contribute to disease—climate (weather, heat, coldness, dampness, dryness), unbalanced diet, nutritional deficiency, nutritional excess, and events and emotions that affect our lives and our bodies. He knew the importance of elimination and emphasized being aware of any differences from normal in the urine or stool.

Hippocrates believed that each patient was unique and that the patient, not the disease, had to be taken care of. He even warned against using the same treatment for the same symptoms in all cases. He used nutrition, fasting, juices, soups, and rest to bring patients with various conditions to a healing crisis.

I often say, "Nature heals, but sometimes it needs a helping hand." I think Hippocrates would have been right at home with that saying. Hippocrates also believed in and loved his work—as I do mine—and that kind of dedication leads to the very best results with patients, I can tell you.

We need to take Hippocrates seriously when he says, "Food should be our medicine, and our medicine should be our food." I certainly do.

"If we eat wrongly,
No doctor can cure us;
If we eat rightly,
No doctor is needed."

—*Victor G. Rocine*
circa 1930

2.

Rocine and His Work

I n the 1930s in Oakland, California, I attended a lecture given by Dr. Victor G. Rocine, a Norwegian homeopath. The lecture deeply impressed me. Our first meeting there set the groundwork for a long-lasting professional relationship and personal friendship.

Dr. Rocine had studied the works of a variety of contemporary European biochemists who had begun to analyze and measure the amounts of the chemical elements in many of our common foods. This growing body of nutritional knowledge had hardly been noticed in the United States at the time. But Rocine had digested it thoroughly.

Rocine proposed that deficiency or excess of any of the primary chemical elements needed in human nutrition was at the root of most human diseases, maladies, and mental problems. In addition, he wrote and lectured extensively on his theory of chemical dominance, that is, the dominance of any particular chemical in a person's makeup—such as calcium, silicon, or sulfur—creates in that person a particular and identifiable temperament type. Though interesting, this is an area of study too extensive for us to give more than a brief mention here.

Like Hippocrates, Rocine believed that food, together with exercise, rest, sunshine, and positive attitudes, is man's best medicine. Rocine became my greatest teacher; he was certainly the man who taught me most of what I know about foods and nutrition. During a period of ten years I studied with him in Oakland, California; Portland, Oregon; and Chicago, Illinois. Rocine once wrote to me in a letter that I was his best student and I felt deeply honored. Because of what I learned from Rocine, I made the decision to focus my sanitarium work on the principles of nutrition.

I was particularly impressed by Rocine's teaching that particular foods have particular effects upon the body. This, in turn, helped me understand how particular foods could be of help in reversing the course of certain diseases. For example, it was obvious, Rocine taught, that in the case of overactivity or underactivity of the thyroid glands, iodine foods would be needed to restore chemical balance in the thyroid gland. Not as obvious, Rocine taught, in cases of joint troubles or digestive system troubles, sodium-rich foods would be needed. Underlining this observation, Rocine stressed an important distinction: much of the population understands sodium to mean table salt. However, this form of sodium is inorganic and useless to the body. The type of salt the body needs is bio-organic sodium salts, which are formed in plants by the internal processes of living cells. Sodium salts are crucial to replenish the sodium used in the joints, stomach, and bowel to neutralize acids and to aid in other metabolic functions. A detailed discussion of the importance of various chemical salts is provided in this chapter.

Hippocrates first proposed that food could be our medicine. With his knowledge of the chemical elements in foods, Rocine furthered food ideas far more broadly. From my training with Rocine, I was inspired to take what I had been given, refine it, and share it with a wider world.

ROCINE—AN UPDATED OVERVIEW

The following selections from the works of V. G. Rocine have been chosen to illustrate the main themes of nutritional wisdom he taught and to show parallels with my own work in nutrition, which is based on many of Rocine's teachings.

CLASSES OF FOODS AND DRINKS

According to Rocine, foods may be classified by function—what they do for us. The following four categories provide an easy-to-understand explanation.

1. Vitality producers, or nerve and brain food.

2. Strength producers, or foods that feed the muscles, ligaments, and bones.
3. Heat producers, or fats and oils.
4. Carriers and eliminators, or juices used in stimulating formation of secretions, digestive juices, or vital fluids at large.

Foods that nourish the brain, nerves, and bone are the phospholipids, phosphate-rich fats and proteins, and sulfur-containing proteins. Foods that supply muscles with strength are the nitrogen-rich proteins. Foods that supply heat are the fats and lipids. Foods that supply energy are the carbohydrates.

A LESSON ON SALTS

The discussion of the function of chemical salts fascinated Rocine. Much of his work is focused on the effects of balancing salts in the body.

Common table salt is a combination of sodium and chloride (a form of chlorine) in equal proportions. It is *inorganic* sodium and chloride. For that reason, it is not as valuable as that sodium which we find in *organic* foods. The human being is organic, or organized. The sodium that we get from plants and vegetables is organized sodium; it is organic. On the other hand, sodium in ordinary table salt, or manufactured salt, is inorganic. When inorganic salt is taken into your system, it will overtax it. As a result, your body will have to work harder in order to throw off the inorganic chemicals introduced. When Rocine spoke about salts, he referred not to the commercial, inorganic variety but, rather, to the organic varieties found in food.

Rocine taught that we must be careful about preserving the chemical salts in foods we eat. This is because when we remove chemical salts from foods, we are likely to alter the other chemicals in those foods. When extracted from food, certain chemical salt may even become poison. For instance, potash (potassium salts, such as potassium carbonate) by itself is a poison, whether it comes from food or from the drug store. This is also the case with phosphorus.

One of the food laws taught by Rocine is that when we are sick we should eat foods that contain the salts made deficient in our bodies prior to or as a result of the sickness.

We should eat food the way God manufactures it for us, Rocine taught. For instance, we should obtain sodium from spinach, strawberries, and carrots. And, when we need an abundance of sodium, such as when a lack of it causes disease in our bodies, we should eat an abundance of those foods that are rich in sodium salts. And we should be careful to eat the proper serving whether well or ill: when we cook spinach and pour its juice into the sink, we are serving the spinach minus the salts. The salts needed were poured down the drain. Similarly, a great deal of sugar and cream on our strawberries may tempt our appetites, but God did not intend for us to eat this way. God did not make sugar. The sugar that people buy in boxes from the grocery store has been refined. God had nothing to do with it. Likewise, when He made wheat, God did not intend for us to eat it in the form of white flour and doughnuts. If God intended for us to eat doughnuts and coffee, He would have made doughnuts and coffee for us. They would have been provided for in the Garden of Eden. Adam and Eve would have had coffee and doughnuts—and just think of where that would leave us!

ACID AND ALKALINE FOODS

Alkaline foods are most valuable for the sick person. Conversely, every food that is acidic, or creates an acidic reaction in the body, is bad for the sick person. (You may wish to consult the Appendix - Food Analysis Chart for a breakdown of alkaline and acid foods.) Every acidic food that generates gas in the alimentary tract is bad for the sick person. Coffee, for instance, may have a good taste, but it makes the stomach acidic. Likewise, tea generates acid and gas— and yet there are people who drink twelve cups a day. Then they wonder why they suffer from nervousness and gas in the stomach!

On the other hand, if you take sodium-containing foods or drinks in abundance when you are sick, your cells become purified. This is because sodium makes the cells alkaline and helps other cleansing elements do their jobs more efficiently. Think of sodium as a power-booster in your body.

THE RELATIONSHIP OF SODIUM, CHLORINE, AND OXYGEN

There is a strong affinity between sodium, chlorine, and oxygen in nature. Foods rich in sodium are often rich in chlorine and oxygen, as well. Oxygen foods are often rich in chlorine, and also in sodium and potassium. Chlorine is the "laundryman" of the body. Sodium neutralizes acids, and chlorine helps to carry impurities off.

Together with sodium, chlorine is used to make commercially-sold soap products. Similarly, a type of "soap manufacturing" is going on in the healthy body. The "soap" manufactured by our bodies by the combining of sodium and chlorine is used to "scrub" fats out of the cells of the body.

SAPONIFICATION

The production of this type of "fat-scrubbing soap" in our bodies is a physiological process called saponification. The process is fueled by the proper quantity and mix of available organic nutrients, including organic salts. In other words, if you take these chemical nutrients out of the body, saponification stops. Without the process of saponification, your body is unable to break down and assimilate fats.

Though this may sound convenient to those who would like to lose weight, stalling saponification does not work to one's advantage. Rather, it will accentuate your present body type. If you are lean, the more fatty food you eat without sufficient organic nutrients, the more lean you will become. Conversely, if you are fleshy, you will grow fleshier without the proper chemical balance. You will swell up with fluids and fatty substances, become food-drunk, and will not be able to get all those fermented substances out of your body without rededication to proper nutrition.

SODIUM BUILDS STRENGTH

The more sodium the tissues take up, the more alkaline and stronger they become. Muscular people who eat sodium-rich foods are the strongest people we have. They are strong because they have a great deal of sodium in their tissues, tendons, ligaments, and joints. All the sodium foods they eat go into the tissues, so that the spleen, the digestive system, digestive juices, and the blood, are robbed of sodium. Thus, their internal organs may suffer from sodium starvation, although their tissues are full of sodium. Muscular women with sodium-rich diets are also powerful, in the majority of cases.

The synovial membrane secretes sodium. If there is a lack of sodium in the joints, the joints crack. Cracking joints show that there is too much calcium, and that there is a lack of sodium in the joints. There is probably a low supply of sodium in the person's diet. Creaking or cracking in the joints is a calcium-excess symptom. Sodium deficiency encourages calcium to emerge from the blood and deposit in the joints.

You can digest albumin well if you have plenty of sodium and chlorine. When you eat the white of an egg and you put plenty of salt on it, you can digest it better. Albumin is usually toxic to sensitive nerves. Nervous people should never eat the white of an egg, nor other foods that are rich in albumin. Tired nerves cannot handle albumin. A man who is pulling heavy loads like a mule can eat albumin foods such as the white of an egg.

If you eat too much calcium in foods, and there is not enough sodium in the system, you are likely to suffer from rheumatism and stiffness.

If you bleed easily, there is not enough of fibrin in the blood circulation system. It means that you are not able to manufacture fibrin because there is a lack of certain chemical salts.

A person who suffers from albuminuria lacks chlorine and sodium. Breathing is difficult without sodium. There is sodium salt in the ear and even in the big toe. Because of the great quantity of calcium in the bones near the ear and big toe, those parts are often calcium deposit sites. Hence, gout begins in the big toe and sometimes in the ear. If you suffer from gout, fall back upon a correct diet. Drink distilled water in abundance, and fall back upon a low calcium diet. Gout is a disease that proves conclusively that there is not enough sodium to keep the calcium in the body in solution.

Lack of sodium results in catarrh, the mucus formed in irritated tissue. If you suffer from catarrh, you need foods that are rich in sodium, such as celery, spinach, and lettuce.

Sodium is important to digestive and eliminative functions. Saliva is rich in sodium. Saliva mixes with food each time we eat. It is important to chew food fully, mixing it with saliva before we swallow, to allow proper digestion. Lack of various types of sodium may cause constipation by limiting the amount of fluid incorporated in stools.

SODIUM DEFICIENCY IS COMMON

People between the ages of seven and twenty-one do not require very much sodium, for between those ages the body is usually alkaline. However, in 99 out of 100 adults,. there is a lack of sodium, rather than an excess.

Both hot and cold climates use up sodium salts in the body, in both cases from perspiration. Being in love enables a person to take up more sodium salts. Under favorable emotions, sodium is more rapidly assimilated or utilized. But under unfavorable emotions of passions (such as fear or anger), sodium salts are eliminated in urine. This is just one more example of how a loving state of mind is favorable to health. Temper, excitement, jealousy, and ugly melancholy passions always have an unfavorable effect upon the chemistry of the body and upon health.

The sick person should have sodium salts or a high-sodium diet. When sodium runs low it signals malnutrition and this malnutrition may occasionally be associated with tuberculosis.

A sodium-deficient person feels well one day, and the next day is sick. He feels that something wrong is going to happen all the time. He is sleepy during the day, and at night his brain is active, very active after midnight. At night he plans and feels as though he can accomplish great things, but during the day he is always sleepy, always tired, always drowsy.

Sodium excess is usually excess of sodium in the tissues, not in the secretions. There is never an excess of sodium in the blood and secretions, but there may be an excess of sodium in the tissues or in the joints.

When sodium is lacking, the nerves are on fire, judgment is unreliable, concentration is poor, and there is a greater tendency to sunstroke or heat stroke.

Sodium is lowered by drinking water, which washes the sodium salts out of the system. People drink too much liquid for their own good. They drink coffee, tea, ice-cold water, chocolate milk, sodas—drinks that not only are not good for the stomach, but are not good for the health in general.

Sodium is lowered through temper and excitement, as well. The more high-tempered you are, the more sodium you burn. At some point, you will suffer from indigestion because of your temper. Again, for this reason, it is much better to cultivate affection than to cultivate temper. If you cultivate love, you will cure your indigestion, but if you cultivate temper, you will fill your system with toxins and impurities.

SALTS IMPORTANT TO PREGNANCY

An expectant mother needs calcium, sodium, iron, and silicon in abundance because her growing unborn baby uses those salts and a great many others.

Morning sickness is an indication of a lack of sodium. Supply sodium in abundance to the pregnant woman, and morning sickness disappears. Chicken- and turkey gizzards contain an abundance of sodium. The pregnant woman suffering from morning sickness can cure it by eating large portions of chicken- and turkey gizzards.

Difficulty in menstruation (or its absence) may be associated with extreme deficiency of sodium and iron salts.

SODIUM TYPES

When there is a lack of sodium, the gastrointestinal system walls become impregnated with calcium salts. To get the calcium out of the walls of the stomach, arteries, and gastrointestinal system, it becomes necessary to eat sodium foods in abundance. The very best way to prevent fatty deposits from accumulating on the walls of arteries is to eat plenty of foods that are rich in sodium.

There are two types of sodium in your body—tissue-bound and free sodium. Tissue sodium is more or less "locked" in the structures; free sodium circulates in the blood and lymph, and is present in the spaces between cells.

Sodium, potash, chlorine, and fluorine are needed when germs are prolific to perform as an antiseptic.

CHLORINE

Chlorine, another organic salt, is found almost everywhere in the body. It is found in the tissues to the extent of 1.75 pounds. Chlorine is needed by the nerves, bones, and tissues; by the sex glands, the sweat glands, and the liver. It is needed in the joints, above all. You can never build strong joints without a heavy chlorine diet, nor can you do it without sodium. You need both of these elements for the building of joints. However, it is chlorine principally that builds joints. Bones cannot be built without chlorine and iron.

If you want healthy children, it is more important to include an abundance of chlorine-containing foods in the diet than almost anything else. You will have healthier offspring if you have all of the salts needed in your body in normal quantities. When you cry, you use up chlorine. Tears are always salty; perspiration is salty.

ALBUMINURIA AND CHLORINE

If starvation of chlorine prevails in your system, you are likely to experience albuminuria—excess protein in the urine. If you find that you have this condition, go to a doctor who understands nutrition and who knows the value of chlorine foods. Inert- and toxic settlements must be cleaned out of the tissues and new nutrients must be supplied. If there is a lack of sodium chloride and potassium chloride, the old settlements cannot be taken out of the body. New nutrients can't be used to rebuild tissue because old material is in the way. This results in tissue congestion. Chlorine is, so to speak, the "laundryman" of the body, as was mentioned previously. It helps remove dead material from the tissues.

If there is an insufficiency of sodium chloride in the system or in the blood, the red corpuscles cannot maintain their shape, nor can they move as fast. If they lack iron, they cannot carry as much oxygen as is necessary. That is the reason that God put important food salts into the body, so that the body may be supplied with oxygen and with all of the powers so essential for self-repair and self-efficiency in all structures and functions. This is also the reason why we should study nutritionally-important chemical elements, both as we find them in the body, and as we find them in food. We should study God's laws of life, and introduce them into our everyday actions, thoughts, and studies.

THE IMPORTANCE OF THE CHEMICAL SALTS

It does not matter so much how much water there is in a certain food, or how much protein, fat, or carbohydrate. Rather, it is exceedingly important to know how much of the various organo-metallic salts you are receiving in your everyday diet.

When the food chemist in his laboratory has burned up a certain food, the ash is left. It is this ash that is very important to study, Rocine noted. This is because the ash contains all of the chemical salts found in the original food. If you wish to cure a sick man, it is the minerals found in the ash that cure. (In speaking of mineral salts, it should be remembered that they appear in the ash as oxides, as sodium oxide, for example.) We should know in which foods we can find those salts. We should know what those salts in foods are good for, so we can supply them when they are needed, especially to those with ailments and diseases.

When we study the chemical composition of carrots, for example, we find that a nutritionist might describe the carrot as having 8 percent silicon. Usually he means the *ash* of the carrot contains 8 percent silicon. To find out the amount of silicon in milligrams, we'd have to make a further calculation.

Suppose we find that goat's milk reduces to 1.74 percent ash, and 31 percent of the ash is chlorine. Since 1.74 percent is the same as 1.74 grams (g.) ash per hundred g. of goat's milk, and 31 percent of that is chlorine, the .31 multiplied by 1.74 is .54 g.. To convert grams to milligrams, we multiply .54 by 1,000 to get 540 milligrams, or mg. That gives us 540 mg. of chlorine per 100 grams of goat's milk.

Vitamins are found in abundance in fresh goat's milk, but in boiled goat's milk many of them are destroyed. Most seeds are rich in vitamins. In fact, any living, growing thing—whether it is a carrot or spinach leaf or chicken—contains vitamins.

No food contains as much of the chemical salts as whey cheese made from goat's milk whey. The second best source is fresh, warm goat's milk. If you can't get fresh goat's milk, you will have to use evaporated goat's milk to get the salts you need. Evaporated goat's milk contains the salts needed by the blood and the heart, as does regular goat cheese. Goat's milk is better for infants than canned milk because it is nearer to human milk in chemical balance.

ALL ABOUT POTASSIUM

Potassium is used by soap manufacturers. The human body is also a soap factory. The function that physiologists call saponification, which we mentioned earlier, means soap-making. It is a physiological process in the human organism. When there is a lack of potassium salts in the body, fats are not converted into the proper compounds. Improperly processed fats may result in a certain form of rheumatism.

The organic food elements are not well taken care of when there is a lack of potassium in the body to neutralize acids. The organic food elements are converted into acids so that the entire body becomes acid. Because of those body acids, some foods will be converted into gases, leading to ailments and diseases. If you do not have enough potassium salts in the body, you had better leave fat alone. Sodium and potassium salts are absolutely imperative in order to take care of fat. Potassium salts are also needed for sugar metabolism.

POTASSIUM AND THE BRAIN

There is one brain center that is essential in order to take care of fatty products. That is the sex brain. If this brain area has become weakened, it is almost impossible for fats and oils to be utilized by the system. Then the spinal marrow, the marrow in the bones, the brain, the lung substance, the joints, and a great many other parts, as well as the secretions of the body, will suffer.

A certain substance called neurolin is needed in the brain and nerves. When the sex brain is weak, neurolin cannot be assimilated nor utilized by the nerves nor by the brain. When the brain is not supplied with this important fat substance, memorization is about impossible, or at least difficult.

A person will experience a peculiar sensation when sodium and potassium are lacking in the system, and when the faculty of affection is weak. When neurolin is lacking, the membranes of the brain cannot function efficiently.

The arachnoid membrane of the spinal cord and of the brain normally secretes a fatty substance needed by the brain, spinal fluid, and male generative fluid. If there is a lack of potassium in the system, this important substance is not secreted.

DEFICIENCY OF POTASSIUM

Deficiency of the potassium element in the body produces certain positive symptoms, and perhaps also disease. How do we know this? If you are a scientific person and want to find out for yourself, cut out from your general diet all foods that are rich in potassium and watch the results in from two to six months. You will then know not only the characteristics, but also the symptoms and ailments peculiar to potassium deficiency.

There is a state of low tissue oxidation, as a general symptom, when potassium is lacking. This means that the tissues cannot function perfectly. Water builds up in the tissues. This accumulation of water indicates the start of kidney problems, and finally the kidneys cannot perform their function. A usual symptom is swollen ankles, which is a sure indication of low tissue oxidation. When

potassium is in low supply, the cerebellum cannot function as efficiently. Hence, the individual feels like falling. Mental work and
physical movements become difficult.

Periodic headache is a symptom that appears regularly, perhaps
at 2 p.m. each day, or in the evening, or perhaps at some other
time. One peculiarity of potassium deficiency is periodicity of
symptoms. Plants, animals, and people who lack potassium present
periodic ailments or symptoms. When there is a lack of tissue
oxidation, the muscles waste away. This is called dropsy. There is
an inward fever, but no outward fever. Something seems to be
burning on the inside, but not on the outside.

The skin itches and the patient scratches, here, there, and
everywhere. The potassium patient complains of sensitive corns.
Self-abuse may be triggered by a lack of potassium. A scar on the
body itches more when there is a lack of potassium.

When there is a lack of potassium, bowel peristalsis is defective.
This may result in acidity in the stomach. Even the muscles become
excessively acid, and soon the whole body becomes acid. This may
lead to many different diseases.

ACID PRODUCES DISEASE

That which we call disease is usually nothing more than the result
of acidity in the system. As long as the system is alkaline, there will
be no disease. But as soon as the system becomes acid, gas is
generated and finds its way into the nerves, and even into the
neurilemma of the nerves, or between the nerve walls and the
nerve itself, giving rise to what is called neuritis. There is no such
thing as an aching nerve, so long as that nerve is alkaline. The
Almighty has given us important foods containing salts of potassium, sodium, silicon, and magnesium to neutralize overacidity. He
knew that without these alkaline elements, our body chemistry
would become abnormally acidic and we would become diseased.

SOURCES OF POTASSIUM

Potassium is found in different greens, vegetables, and cereals.
Potassium is usually found on the outside, but it is also functioning
inside, the cells and in the inner sanctuary of the tissues. The
manufacturers mill the exterior from cereal grains, and most of the
potassium, silicon, phosphorus, and other important organo-me-

tallic salts (as well as vitamins) are milled off. These cereal makers sell us only the starch.

CHEMICAL ELEMENTS ARE VITALLY IMPORTANT

Among the chemical elements, magnesium is most essential for the nerves and brain. Silicon acts on solid tissues, and also upon the skin and hair. Silicon helps keep the more solid tissues alkaline. Potassium makes the muscles and vital organs more alkaline. Magnesium acts upon phosphorous products. Sodium keeps calcium in solution and acts upon the blood and digestive juices.

After the tissues become acid, we get sick, and no doctor can cure us with drugs or surgery. You may go to a hospital and be operated on, but when you come out you are as acid as when you went in. You have just as much acid in your body as ever. If you do not fall back on sodium, silicon, potassium, and magnesium, the Almighty's remedies which He puts into foods to make the tissues alkaline, it is impossible to get well. So long as the tissues are alkaline, we can keep well.

Chemical elements aid in regeneration of the species. There is nothing better for impotence than oatmeal, raw egg yolk, iron, phosphorus, and sleep. Anyone suffering from sexual weakness will become more invigorated by such a regimen. If a husband or wife would fall back on an oatmeal diet, he or she would more able to retain the spouse's affection

Chemical elements play a role in nervous system function from muscle ticks to mental clarity. A neurosis can be a nerve habit, as, for instance, the "writer's cramp." In such an ailment, silicon, sulfur, and neurolin are used up in the nerves. Neurosis is in the nerves. In times of nervous debility, fall back on oatmeal. Cook the oatmeal slowly for a long time. Eat oatmeal muffins. They are essential in an alkaline diet. So long as the tissues are acid, so long as there is acidity in the secretions, so long as the sexual system is acid, we will be weak and sickly.

Silicon has a remarkable effect upon the cerebral cortex, or that part of the brain that thinks and observes. Take out that part of the brain and we are as helpless as babies, yet we may have as strong a will as we ever had. Then we are human beings without intellect. We are next to animals.

Silicon is beneficial for the nerves, the bowel walls, the alimentary tract, the lungs, and the entire system. If children are not supplied with proper amounts of silicon, they suffer from catarrh.

Catarrh develops in membranous tissue, or in the walls of the gastrointestinal system. If we supply silicon in abundance, we become more alkaline.

THE SILICON PERSONALITY

A silicon person is here, there, and everywhere. He is like a goat. He likes to be high in the air. The goat likes to get up among the highest rocks. Not surprisingly, goats are partial to silicon-containing feed, grasses, leaves, and grains.

Silicon people are elastic and prone to exaggerate. They see everything through magnifying glasses (imagination). They are busy all the time. When they are not talking, they are singing. But they never talk long on any one subject. You find silicon people mostly among the Swiss. Swiss people have beautiful hair.

There are two elements that make us feel important: one is silicon, the other is phosphorus. If there is an excess of phosphorus, a man feels like a king. He walks with a stately bearing. A woman feels like a queen. The difference between silicon people and phosphorus people is that silicon people talk, while phosphorus people stand and look. The silicon man talks all the time, looking this way and that way, in every direction. He talks on a thousand different subjects. He talks and gestures as actively as a goat.

The silicon person is sociable, just like the goat. If you buy a goat, be sure you buy two. If you do not buy two, you will lose the one you buy. That is the way of goats. Silicon persons are the same. They can never be by themselves.

Silicon persons do not need much sleep. They may retire at 1 a.m. and sleep until 4 a.m. They are like electric engines—full of activity. Silicon people never seem to live to be very old. They become so excessively active that they simply wear themselves out.

EFFECTS OF SILICON DEPLETION

Silicon-deficient patients think they are going to die. They cannot bear noise. When they are tired, they cannot get hold of their thoughts; they cannot form ideas; they cannot recall subjects. They feel as if they are losing their minds. They may have everything they could wish for, but they think they have nothing.

The person who suffers from silicon-hunger is in the same kind of condition as an electrical motor that has run itself out. Now he sleeps so you cannot wake him, later he cannot sleep. When he tries to talk to you, he tires quickly. He believes there is some disease at work in his system.

Silicon cannot be massaged into a patient, but a silicon diet together with massage treatment will yield beneficial results. A silicon-deficient person always staggers to the right side, never to the left.

Silicon is needed when the orifices of the body become narrowed or closed-up. When a patient tells you that his arm is upstairs and his feet are down in the basement, he is suffering from silicon hunger. Silicon is needed when a patient craves stimulants.

It is the function of silicon to supply cerebellar energy. If you feel exhausted or overworked, fall back on silicon. It does not feed your brain or nerves, but it stimulates them. That is the way silicon acts. It tones up the cerebellum and the motor nerves. When there is a normal amount of silicon in the system, the brain and the nerves are stronger. Increase the silicon, and the nerves become strong. But, if you do not work, you cannot assimilate silicon. Silicon requires motion. When we labor, we take up a greater amount of silicon. But if we sit around, ride in automobiles, and never walk, we do not take up silicon and we grow weaker.

To increase silicon in the body, leave fats, starches, and sugars alone. Eat plenty of oats and oatmeal and you will grow stronger in your body. People in Southern California, where there is sandy soil, a high altitude, and a dry climate, are successful in raising silicon-loving goats. This is because everything there favors silicon metabolism—both in goats and in people. People in a cold climate are often disagreeable without adequate silicon. But in California, where it is dry and warm, and the functions of the people are more alkaline, the disposition becomes more genial.

SILICON'S ROLE IN NOSE-BRAIN COMMUNICATION

If the frontal sinuses become catarrhal, you suffer because there should be a free outlet through the frontal sinuses. There is a sieve-like bone at the root of the nose, leading up into the brain, called the ethmoid bone; if this is filled up with catarrhal mucus, your thinking, memory, concentration, and health are all affected unfavorably.

The nose has another function besides that of inhaling air for the lungs, and that is the function of communication with the brain.

If the frontal sinuses are closed, there will be a multiplicity of ailments. When there is a lack of silicon, these nasal canals become catarrhal and are filled up with mucus or phlegm, resulting in sickness. Our memory and faculty of understanding and other mental faculties are inefficient. A certain amount of evaporative moisture passes from the brain through the nose every hour. If the nasal canals and the ethmoid bone are congested, the brain avenues of elimination are closed, resulting in colds, catarrh, disturbances of the brain, headache, and other ailments.

MAGNESIUM—THE CALMING ELEMENT

Children who scream are using up magnesium in the system; when babies scream so that you have to walk them all night, they lack magnesium.

Nursing mothers of these babies should eat foods rich in magnesium. The needed magnesium will be supplied in the breast milk. Another thing, that little fellow should be kept awake between 5 p.m. and 10 p.m. If he is kept awake all that time, he will sleep later on. Wear him out, so he can fall asleep at 10 p.m. If you do not do that and do not supply magnesium, he will grow weak crying. The parents will become nervous and irritable taking care of him. They will burn up their magnesium, their complexions will fade, and their eyes will lose their brightness—all because of a lack of magnesium.

PHOSPHORUS—A VITAL BRAIN AND NERVE ELEMENT

When the heat in the brain becomes excessive, it affects the function of the brain. If a fever rises to between 106 and 108° F, phosphorus in the brain can begin to melt. In addition, when there is an excess of heat in the brain, there is weakening of the brain functions. The physical functions of the body are transmuted, or altered. You may grow weak and irritable. You tremble, and your nerves tremble. Be careful in regard to artificial heat so that it does not become excessive. Never sit in an overheated room. The brain may become overheated.

The head absorbs heat and light more quickly than other parts of the body. Be sure to cover up your head in intense light and heat, for the characteristics of phosphorus are such that it absorbs the hot sun's rays easily. It is better to go into the sun with a nude body than with an uncovered head. If you suffer from the sun's rays, when the temperature is as high as 105° F, be sure to cover your head, or place green leaves or green grass in your hat. As long as you protect your head, you are safe from too much heat. If you have a patient suffering from excessive fever, protect his head and you will save your patient. So as soon as the fever heat in the body runs above 105° F, there is great danger.

EMOTIONAL PEOPLE MUST GUARD THEIR DIETS

There are dietary precautions that should be observed by people with prominent and persistent temperament characteristics. Emotional people, for example, should avoid foods with malic acid.

Three or four apple varieties are comparatively free of malic acid. The Astrachan, Belleflower, Jonathan, and Delicious are almost entirely free from malic acid. Emotionally-sensitive persons should eat these types of apples to remain healthy.

In addition, people of this temperament should become acquainted with all sorts of fruits that do not carry malic acid or tartaric acid. Tartaric acid foods and malic acid foods are very destructive to the nerves of emotionally-sensitive persons. If they eat fruit containing these acids, they may develop various kinds of nervous diseases.

CURING CRIME WITH NUTRITION

Rocine believed that if we had a sanitarium for hardened criminals, where they could associate with the most perfectly serene and kindly religious people, and also eat a balanced diet, they would develop the higher religious sentiments. He believed that we could cure criminals more successfully through right diet and right associations than by simply condemning them to "serve their time." Rocine believed that as long as we have criminals, we cannot uplift civilization. So as long as we let the criminal eat any kind of food, especially the kind he likes to eat, we cannot expect that poor person to break the criminal habit. He simply feeds the lower faculties; he lives, so to speak, in hell already.

PREPARATION AND COMBINATION OF FOODS

Foods should be cooked in a fireless cooker or broiled over a clear, hot fire, free from smoke. Soup should be cooked slowly. Mutton and fish can be boiled in hard water; the scum from cooking should be removed. Vegetables may be cooked in soft water. Cabbage, carrots, rice, and oatmeal can never be boiled too long. Avoid frying foods or quickly boiling soup, since rapid heating hardens the fibers; salting the meat during cooking causes the fluids and meat salts to be lost. Boiling spinach extracts the most valuable salts into the water.

For a detailed analysis of the alkaline and acid content in foods, refer to the chart in Chapter Three. In addition, Part Two of this book addresses other specifics about food combining and nutritional content.

Some fruits are acid, but most are alkaline. Fruit is best eaten by itself between meals; vegetables and grains by themselves; beans and peas by themselves in small quantities. A weak stomach can digest fruits more easily in the morning and evening. Fruits should be eaten at the beginning of a meal, not after.

Breakfast should consist of fruits or vegetables, and milk toast; or apples, cherries, or berries. Avoid soft, mushy foods; soft breads; and pancakes.

At the noon meal, eat vegetables and, possibly, meat.

At the evening meal, drink milk, eat cereal or fruit.

FOOD CHOICES AND TIPS

As soon as the milk is drawn from the cow or goat, its vitality commences to disappear and passes into the air. Rocine observed there is a similar life principle in every food and every living thing. When it departs, the food is of no value nutritionally. As soon as fruit is picked from the tree, the life principle begins to depart and goes back into the atmosphere, leaving the dead, organic material behind. Honey is valuable not because it is sweet, but because it retains its life principle longer.

Fruits, berries, greens, and vegetables contain more waste bulk than nutritionally-valuable substance. If eaten alone they produce diarrhea and debility, but if taken in connection with more nutritious foods, they keep the bowels active, and the system cool and free from a surplus of more stimulating food.

Hot foods and drinks should not be taken because they weaken the mucous membranes and result in spongy gums, decayed teeth, and swelling of the mucous glands of the stomach. They weaken the stomach.

Cold foods and drinks absorb the heat of the stomach, arrest digestion, and produce irritation. Foods must be heated in the stomach. If they are cold, this weakens the stomach and weakens the brain center that generates nerve force to the stomach. To eat and drink ice-cold or very hot drinks and foods is to weaken the faculty of affection. The temperature of the stomach should never be lowered by cold foods and drinks, nor should foods and drinks be too hot. It is not well to eat when the body temperature is low. This lowers the digestive power.

Foods difficult to digest are better nutritionally because they have staying power. They are best for people who exercise and do physical work. But such foods will not do for a delicate stomach or for people with sedentary occupations and habits.

WHAT YOU SHOULD KNOW ABOUT MEAT

Meat is rich in protein. Fat meat furnishes heat and energy, and lean meat builds muscle. The dark of any meat is rich in phosphates and in muscle. But the white meat comes from a part of the animal that is not exercised. Therefore, it contains no more nutrition than white bread.

Humans must cook meat to purify it for digestion. The cooking process converts into gluten the collagen between the meat fibers. This allows the digestive juice called pepsin to act upon the fibers more readily. This is an advantage in digestion, but cooking also coagulates the proteins, which is a disadvantage. Thoroughly pounded and minced meat is more easily digested than either cooked or fried meat.

Raw meat juice contains bile salts needed by the system. The salty elements in meat are best extracted by roasting, or by making a gravy of the drippings of roasted meat, the fat removed. Lean mutton, roasted pigeon, turkey, duck, and grouse are rich in these elements or salts. The best meats are from animals raised on vegetables. In addition to those mentioned above, other good meats are veal, duck, prairie chicken, game, and goose. Many other meats actually contain poisons. For example, meat from animals such as the pig is full of ferments, uric acid, and xanthin—all poisons. It is not pure meat.

LET'S GET OUR MINERALS FROM FOODS

We can find all the elements necessary for health in our foods. If we need silicon, we find it in oats and barley. If we need calcium, we find it in milk and cheese. If we need chlorine, we find it in sauerkraut. We find sulfur in all cabbages. Why should we go to mineral springs for these elements? We can find them in foods—and in organic form, not inorganic as they are in mineral springs.

To cure the patient, we must find the cause and change the diet. But we cannot cure a person while he is filled with impurities. We must eliminate those impurities before the healing process can begin.

Perhaps it is the work the patient is doing that makes him sick. If he is a bookkeeper and sits all day, the mathematical centers in the brain become weak because he overworks those centers.

When the vital centers of the brain become congested, there is a lack of nerve force. No energy is transmitted. The mechanism is weakened. We need all the knowledge we can get in order to cure people. We need knowledge of diet, and still more. We never can know too much when it is a question of health.

ROCINE IN PERSPECTIVE

Rocine came close to perfecting what Hippocrates proposed 2,000 years before.

Although Hippocrates knew that food could be our medicine, this chapter illustrates how Rocine developed food ideas far more broadly. From my own training with Rocine, I was inspired to take what I had been given, refine it as much as I have been able, and then share it with a modern world.

In the next chapter, as in my work, I combine a discussion of my own nutritional research with the works of both Hippocrates and Rocine.

"Nature cures
when given the opportunity."

—Bernard Jensen
1988

3.

My Work in the Health Arts

Hippocrates was a physician who looked to nature for solutions to health problems. He looked to the "dust of the earth," the minerals assimilated by plant life. He went to fruits, berries, and so forth and found that they had effects upon the body. My own experience has taught me that some foods have quite remarkable effects. However, the average nutritionist doesn't realize the power and the energy that are available from foods. Hippocrates started me thinking that foods generate power, health, and a wonderful sense of well being. I'm glad he did.

V. G. Rocine helped me understand the link between chemical deficiencies or excesses and ill health, and his teaching inspired me to take up nutrition as my life's work. Dr. Rocine taught that foods had power and that foods transfer their power to human beings when we digest and assimilate them.

In this chapter, I discuss my own philosophy of natural, right living, and foods that heal. If you notice parallels between information in this chapter and that covered in previous sections of the book, you're paying attention. It's no mistake that I have chosen to build on the teachings of Hippocrates and Rocine. In fact, they have been my inspiration to take the study of nutrition further into our modern age.

THE NEGLECTED STORY OF FOOD CHEMICALS

One of the basic premises behind foods' ability to heal is that they
are formed of organic chemicals, just as we are. From the soil come
the chemical elements. As the Bible says, we have been fashioned
from "the dust of the earth." The only chemical difference between
soil and human beings is that the chemical elements in our bodies
are structured in a more highly-evolved order than they are found
in the topsoil of this planet. However, humans cannot assimilate or
metabolize chemical elements from soil. Nature has dictated that
we must obtain them from animals and plants.

At the lowest rung of our ecological food chain are found the
green plants. Plants take up the chemical elements dissolved by the
moisture in the earth's soil. Their minute root hairs draw water
particles containing exact chemical elements necessary to manufacture a parsley stem, or a spinach leaf, or a strawberry. It has been
estimated that the combined total length of the roots of one oat
plant would equal almost 3,000 miles. Think of the amount of
nutrition a plant with that type of root system will draw from the
soil!

Likewise, the alfalfa plant has massive roots for its size—traveling as deep as 250 feet into the earth to draw up all of its needed
nutrients. As a result, the alfalfa plant is rich in silicon and many
mineral salts stored in the earth's deep recesses.

THE ECOLOGICAL FOOD CHAIN

Humans are at the highest rung of the ecological food chain. We are
able to assimilate nutrients by eating all the other various partici-
pants in our food chain: plants, plant-eating animals, and animals
that eat plant-eating animals.

Recent medical research has pointed out that the more fresh,
unprocessed foods we eat, and the "lower" on the ecological food
chain we eat (i.e., the more plants and plant-eating animals we
eat), the better chance we have of maintaining our health and
guarding against disease. This is because impurities are carried
through the food chain. Therefore, for argument's sake, if cattle
feed is sprayed with insecticide and we eat the beef from the cow
that ate that feed, we will also be eating the harmful chemicals that
settled in to the tissues of that cow.

On the other hand, if we eat foods that are grown in organic
soil without harmful pesticides, we are eating pure foods. If we eat

animals that eat organically-grown plants, such as wild game or organically-fed livestock, we are reducing our chances of ingesting harmful substances through the food chain. In other words, we are eating "low" on the food chain.

Once we have chosen the freshest, most-wholesome types of nutritionally-sound foods, we must eat them in the right proportions to maintain health. We must educate ourselves on the effects each food has on our chemical makeup, on our metabolism, to maintain health. Hippocrates observed and recorded the types of changes that different foods brought about. This is why he called food a medicine. Medicines, like foods, have an effect. Unlike medicine, food enables us to build up or tear down the actual tissues of our bodies. When we eat right, we build strong, healthy bodies. But when our diets are defective and our lifestyles are degenerative, we tear down our bodies by destroying vital tissues.

Like Hippocrates, I have observed that when people omit certain foods from their daily regimen, health problems almost certainly develop. Deficiencies cause problems. This is why I speak of dietary deficiencies as "sins of omission." They are far more dangerous than "sins of commission." This is quite important to comprehend: it is not always what we eat that counts. More importantly, it can be what we do not eat that causes us problems with our health and, indeed, our happiness.

THE FOUR MOST COMMONLY DEFICIENT ELEMENTS

The four chemical elements most frequently deficient in my patients have been calcium, iodine, silicon, and sodium. Foods that are highest in these minerals can be easily integrated into any meal plan. I believe most people are slightly-to-seriously deficient in these chemical elements and that many unnecessary health problems occur as a result.

Calcium

It is important to note at the beginning that up to 32 percent of available calcium is destroyed when food is heated above 150° F (the boiling point is 212° F). Consequently, pasteurized milk is a limited source of calcium, as are canned foods, which are processed using excessively high heat.

We eat plants.

We consume meat and milk from livestock.

Pesticides used on plants are ingested by livestock when they eat those plants.

Livestock eat plants (crops).

The Food Chain

One of the highest calcium broths, especially for growing children, is made from barley and green kale. This is an old Danish soup very good for growing children. Cereals and grains are also excellent calcium sources. A concentrated supplemental form of calcium is bone meal or, for vegetarians, calcium lactate. Natural calcium supplements (in food form) will not cause hardening of the arteries or calcium deposits, affect blood pressure, or produce other adverse symptoms. Dolomite is an inferior calcium source because the body's ability to assimilate it is questionable.

High sources of food calcium are sesame seeds, dulse, Irish moss, kelp, and greens. Seeds, nuts, and cereal grains (unrefined) are all excellent sources of calcium. High calcium foods are listed below (* denotes highest sources).

Agar
Almonds
Avocados
Barley
Beet greens
Beans
Blackstrap molasses
Bran
Brazil nuts
Broccoli
Brown rice
Brussels sprouts
Buckwheat
Butter, raw
Cabbage
Carrots
Cauliflower
*Cheeses: hard and
   cottage; raw
Chia seeds
Coconut
Cornmeal, yellow
Cream, raw
Dandelion greens
*Dulse

Egg yolk
Figs
Filberts
Fish
Gelatin
*Greens
*Irish moss
*Kelp
Kohlrabi
Lentils
*Milk: goat and
   cow; raw
Millet
Oats
Onions
Parsnips
Prunes
Rice polish
Rye
*Sesame seeds
Soy milk
Veal joint jelly or
   broth
Walnuts
Watercress
Wheat (whole)

Iodine

Kelp and dulse (Nova Scotia) head the list of iodine-rich foods. Others are listed below.

Agar	Mustard greens
Artichokes	Oats, steel cut
Asparagus	Okra
Bass	Onions, green and dried
Beans	Oysters, raw
Blueberries	Peanuts
Brussel sprouts	Perch
Cardamom	Pike
Carrots	Potatoes, sweet and white
Chervil	Quail
Chives	Rutabaga
Coconut	Salsify
Cucumber	Seaweed
Eggplant	Silver salmon
Fish	Sole
Fish roe	Spinach
Garlic	Squash
Goat cottage cheese	Strawberries
Goat milk, whey	Swiss chard
Green peppers	Tofu
Green turtle	Tomato, ripe
Haddock	Trout
Halibut	Tuna
Herring	Turnips
Kale	Turnip greens
Leaf lettuce	Watercress
Loganberries	Watermelon

Silicon

Oats and barley head the list of silicon foods. More concentrated supplemental sources are rice polishings, rice bran syrup, kelp, oat straw tea, and alfalfa tablets. These and other good food sources are listed below.

It is important to note that all refined starches and carbohydrates (i.e., white flour, white sugar, white rice) are nearly devoid of silicon because the outer skins and hulls, which contain silicon, have been removed in the refining process. Unrefined nuts, seeds, grains, and cereals have a generous supply of silicon in their outer coverings.

Alfalfa broth and tea	Lettuce
Apples	Marjoram
Apricots	Millet
Asparagus	Nectarines
Bananas	Oats
Barley	Onions
Beans	Parsnips
Beets	Plums
Beet greens	Potatoes, sweet
Cabbage	Pumpkin
Carrots	Raisins
Cauliflower	Rice, brown and wild
Celery	Rice bran and syrup
Cherries	Rice polishings
Corn	Spinach
Cucumbers	Sprouted seeds
Dandelion greens	Strawberries
Dates	Sunflower seeds
Figs, dried	Tomatoes, ripe
Grains	Turnips
Greens, mustard and turnip	Watermelon
Horseradish	Wheat bran
Kelp	Wheat germ
Kohlrabi	Whole wheat

Sodium

Veal joint broth and powdered whey (cow's milk or goat's milk) are highly-concentrated sources of sodium. Goat's milk or whey and black mission figs are a superior sodium combination (and this is also a champion arthritis remedy).

Fair sources of sodium include other cabbage, water chestnuts, garlic, peaches (dried), radishes, broccoli, Brussels sprouts, and cashews. High sodium foods include:

Apples
Apricots, dried
Asparagus
Barley
Beets and greens
Cabbage, red
Carrots
Celery
Cheeses
Chick-peas, dried
Coconut
Collard greens
Dandelion greens
Dates
Dulse
Egg yolks
Figs
Fish
Goat milk
Horseradish
Irish moss

Kale
Kelp
Lentils
Milk, raw
Mustard greens
Okra
Olives, black
Parsley
Peas, dried
Peppers, hot red and dried
Prunes
Raisins
Sesame seeds
Spinach
Strawberries
Sunflower seeds
Swiss chard
Turnips
Veal joint broth
Whey

CHEMICALS IN FOODS HAVE SPECIFIC JOBS

Each chemical element has its own unique vibratory rate or electromagnetic field that helps it to do a specific job. When these elements are combined in a certain physical structure, they form the organs of the body. When they are combined in terms of their activities, they make up the cellular functions at one level and the particular organ functions at another level. If we accept that chemical elements are necessary for health, we must also accept that a lack of the proper quantity of chemicals can lead to health problems. In the human body, science has determined that chemical deficiencies affect not only the physical *structure* of tissue, but also its *functional* activities. In other words, without proper potassium supply the heart tissue will not be as healthy, and this will impair the heart's ability to function.

Some organs and bodily structures are more seriously affected by deficiencies in one chemical element than others, because some organs act as special repositories or storehouses for those elements. For example, the thyroid gland stores iodine and the fingernails, hair, and skin store silicon. Therefore we assume if iodine or silicon are lacking in the respective organs in which they are stored, the same elements are almost certainly deficient in the rest of the body. The rest of the body has used up the reserves at this point.

When one vitamin is depleted, the others can't be far behind. For example, vitamin B cannot be held in the body unless there is enough silicon to hold it until it is used. Other vitamins are similarly linked to the chemical foundations established to hold them until needed.

When we are using a balanced food regimen, these chemical elements are being replaced as rapidly as they are being utilized. When we lack any of the chemical elements, disease may develop and symptoms appear. Several follow.

In psoriasis, a skin disease, there is a lack of silicon, just as we find in boils. We lack silicon whenever there is excess mucus in the body. We lack silicon where cysts or growths are present.

Stomach ulcers indicate a lack of sodium. Calcium deficiency contributes to osteoporosis. Without adequate phosphorus or sulfur, brain function would be impaired. Fluorine, a very unstable element, can give teeth and bones stability and hardness by reinforcing the calcium.

Each of these chemical elements comes to us from the earth, is assimilated into foods, and must be eaten in the proper amounts by humans in order for us to function properly.

THE BASIS OF GOOD NUTRITION

The basis for proper nutrition is found in whole, pure, and natural foods. The number of calories in a meal means nothing unless they come from a proper balance of foods. If there is inadequate protein, the diet can cause you to become ill. If there is inadequate vitamin and mineral content, the same may happen. Even fats, in small quantities, are necessary for metabolism. Without the proper balance of a variety of nutrient-rich foods on a continuing basis, good health cannot be achieved and maintained.

THE FOOD LAWS

If you don't know the basic food laws for good nutrition that I have discovered and applied to eating, you should learn them. I have used them for my own patients. I have noted that the proportions have to be a certain way in the body. Every day, we should eat servings of six vegetables, two fruits, one starch, and one good protein. For suggestions for each, you may refer to Appendix A. In addition, you may refer to my Food Laws diagrams on page 49. When these proportions are adhered to, you will have achieved an 80 percent alkaline/20 percent acid diet. Your 20 percent acid comes from your acid starches, and proteins. Your 80 percent alkaline comes from your six vegetables and two fruits.

An easy-to-follow method to achieve this balance is to pair one starch with several vegetables at one meal; one protein with several vegetables at another meal; and to have fruits as in-between-meal snacks. Of course other possibilities are numerous and, after you begin eating in these proportions, you'll surely develop meals that suit your family's tastes and lifestyle best.

This 80/20 ratio of alkaline to acid also is present in your blood, which contains 80 percent alkaline nutrients to 20 percent acid nutrients. I don't know whether Hippocrates knew this. I am certain he did know that, when you eat the proper proportions of foods, your body balances itself.

The same thing happens in the soil. There is a God-given, innate property in plants that reaches out and takes what it needs from the soil. Some plants are richer in certain chemicals than others. And, for this reason, you must have variety; you must have the right proportions, and you must have 60 percent raw foods each day. Raw foods are important because they contain all the nutrients, fiber, and enzymes that specific food has to offer. Cooked, canned, or processed foods, on the other hand, lose nutrition. Of course some foods, such as meats and fish, must be cooked to kill harmful bacteria. You must have a certain amount of protein daily and a certain amount of starch. These, too, should be varied from day to day. You must know your proteins and starches. In fact, you should know what all foods should do for you. They have definite functions in the body.

ACID AND ALKALINE FOODS

There has been much talk in the last few years about acid and alkaline foods. It has been determined that for optimum health, a body should be slightly alkaline. If the body becomes too acid or too alkaline, illness and disease will result.

When we speak of an acid body, we need to understand the notion of pH, or relative acidity-alkalinity, in the standard terminology of chemistry. The pH scale is from 1 to 14, with 7 represent-

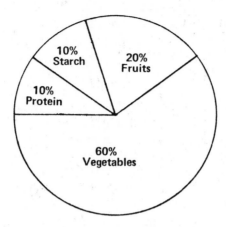

Food Proportions

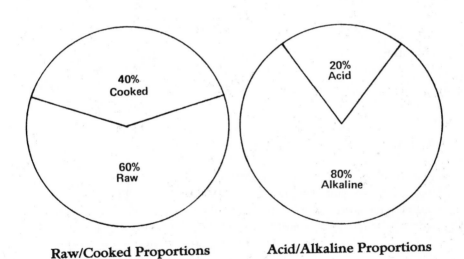

Raw/Cooked Proportions   **Acid/Alkaline Proportions**

The Food Laws

By choosing to eat 60 percent raw foods in the proportions shown
here, we will be creating in our bodies an 80 percent alkaline/20
percent acid ratio that is super nutritious and safeguards against
disease.

ing neutrality, neither acid nor alkaline. Most living things require an acid-alkaline environment of between pH 5.5 and 8 to survive. Outside this range, they sicken and die.

As a living plant or animal approaches the outer limits of its pH tolerance toward either the alkaline or acid side, tissue damage and loss of function begin to occur. Of course, in animals or people, some organs such as the stomach and kidneys maintain a higher acidity than other organs and tissues.

The optimum pH range for human tissue is about 7.35 to 7.45, which is very close to neutral, slightly favoring the alkaline side. This is the normal pH for human blood plasma.

When we use an excessively acid diet (meat and potatoes every day, for example), large quantities of sodium, potassium, calcium, magnesium, and other acid-neutralizing chemicals are constantly used up in preventing tissue damage. The problem is, these chemical elements are drawn from the tissues and must be replaced, or deficiencies will occur, soon followed by disease symptoms. Over alkaline diets are seldom found, but they would also be harmful over the long term.

For these reasons, my food laws say we must have six vegetables and two fruits each day, which make up the 80 percent of the alkaline-forming foods required, and one starch and one protein, the 20 percent acid-forming foods required. Keep in mind that most foods are very close to the neutral pH point of 7, slightly acid or slightly alkaline.

ACIDITY AND ALKALINITY IN THE BODY

The following experiments illustrate the point: medical researchers Balint and Weiss took a rabbit and alkalinized it. Then they slit the rabbit's leg and injected turpentine. The damage to the leg was very slight. When the rabbit was acidized, however, the injection of turpentine was followed by inflammation, tissue sloughing, and death.

In another experiment, this time with scarlet fever patients, it was discovered that when urinary acidity ranged high, there was an aftermath of kidney disease in two-thirds of the cases. When the acidity was low, only 3 percent of the cases were thus complicated. These experiments point out very definitely that a high alkaline balance is the body's first line of defense against illness and death. Refer to Appendix A—Food Analysis Chart, to help you choose the most nutritionally-balanced foods for you and your family.

In addition to the proper acid/alkaline balance, we must supply our bodies with specific nutritional requirements. Problems develop when we lack nutrients to build the muscles, to build the ligaments, to build any part of our bodies. When a certain organ is not healthy, we should be able to go to nutrition and bring back the basic building blocks it needs. Our organs are designed to be self-repairing, but they need an opportunity to do this.

THE BODY MOLDS TO WHAT WE FEED IT

The body changes according to what we eat. It will mold to a salad, it will mold to junk foods—sugary foods, fatty foods—it will mold to white vinegar, wine vinegar, or to a natural apple cider vinegar. The bad news is that your body molds to these things, and it does the best it can with the foods that you give it. The good news is that when you change your diet to a proper, balanced way of eating, your whole body changes with it.

HEALTHFUL JUICE COMBINATIONS

The following specific disorders of the body can be aided by using juices as directed. (For juice combinations to boost the health of specific systems, refer to the chart beginning below.) Nearly every disorder responds well to appropriate vegetable juices. Celery, parsley, and carrot juices are good for any condition. They can be mixed or can be used separately. The juices suggested for a specific bodily disorder may be used alone or in combination with others.

If you bathe the tissues of your body by taking these juices, you will in time cleanse them and then rebuild, rejuvenate, and feed a starved body. Don't get the idea that you should live on them entirely. Take them with meals and between meals. For the best results, drink at least one pint a day. Note: Drink these juices with a straw, or sip slowly, to allow them to mix well with saliva.

Health Cocktails for Common Disorders

Disorder	Health Cocktail
Anemia	Blackberry and parsley juice; parsley and grape juice.
Arthritis	Celery and parsley juice.

Health Cocktails for Common Disorders—*Continued*

Disorder	Health Cocktail
Asthma	Celery and papaya juice; celery, endive, and carrot juice.
Bedwetting	Celery and parsley juice.
Bladder ailments	Celery and pomegranate juice. (Pomegranate juice is the best for the bladder.) Also good, shavegrass herb tea.
Blood ailments	Blackberry juice, black cherry juice, parsley juice, dandelion juice. Tomato juice and desiccated liver.
Blood pressure (high)	Carrot, parsley, and celery juice; lime juice and whey powder; grape juice and carrot juice.
Blood pressure (low)	Parsley juice, also capsicum and garlic.
Bronchitis	Juice of 2 lemons, 3 T. honey to one pint of flaxseed tea. Use one tsp. every hour. Or bake a lemon, juice half of it and add to one cup of oat straw or boneset tea. Then, go to bed and perspire.
Catarrh, colds, Sore throat	Watercress and apple juice with 1/4 tsp. pure cream of tartar.
Circulation (poor)	Beet and blackberry juice; parsley and alfalfa juice with pineapple juice; grape juice with one egg yolk.
Colds and sinus	Celery and grapefruit juice; watercress and apple juice with 1/4 tsp. pure cream of tartar; coconut milk and carrot juice; celery and grapefruit juice with 1/4 tsp. cream of tartar.
Colitis, gastritis, gas	Coconut milk and carrot juice.
Complexion (yellow)	Grapefruit juice.
Complexion problems	Cucumber, endive, and pineapple juice; one T. apple concentrate; 1/2 glass cucumber juice and 1/2 glass water.
Constipation, stomach ulcers	Celery with a little sweet cream; spinach and grape-fruit.

Disorder	Health Cocktail
Diarrhea, infection	Carrot and blackberry juice.
Eczema, scurvy	Carrot, celery, and lemon juice.
Fever, gout, arthritis	Celery and parsley juice.
Gall bladder	Radish, prune, black cherry, and celery juice; carrot, beetroot, and cucumber juice; prune, black cherry, celery, and radish juice.
Gallstones	Beetroot and radish juice; green vegetable juices.
Glands (for building)	Pineapple juice with one egg yolk, one T. wheat germ, $1/4$ tsp. powdered Nova Scotia dulse—take daily between meals; $3/4$ cup carrot juice, $1/4$ cup coconut milk, one T. wheat germ, one tsp. rice polishings or rice bran syrup, 1 cup tomato juice, one T. cod roe.
Glands and nerves	One T. cherry concentrate, one tsp. chlorophyll, and one egg yolk.
Glands, goiter, impotence	Celery juice, one tsp. wheat germ, and one tsp. Nova Scotia dulse.
General house cleaning	Celery, parsley, spinach, and carrot juice.
Gout	Celery juice; combination of celery and parsley juice.
Heart	Carrot and pineapple juice with honey; liquid chlorophyll (alfalfa); parsley, alfalfa, and pineapple juice.
Hair (to improve)	One T. cherry concentrate, one tsp. oat straw tea to a cup of boiling water. Steep tea 10 min., then add cherry concentrate.
Indigestion, underweight	Coconut milk, fig juice, parsley, and carrot juice.
Infections	Carrot and blackberry juice.

Health Cocktails for Common Disorders—*Continued*

Disorder	Health Cocktail
Insomnia (sleeplessness)	Lettuce and celery juice.
Jaundice	Tomato and sauerkraut juice, one glass every day for a week.
Kidneys	Celery, parsley, and asparagus juice; carrot and parsley juice.
Kidneys (bladder) problems	Black currant juice with juniper berry tea; pomegranate juice and goat whey; celery and pomegranate juice.
Liver	Radish and pineapple juice; black cherry concentrate and chlorophyll; carrot, beet, and cucumber juice.
Memory (poor)	Celery, carrot, and prune juice and rice polishings.
Nervous tension	Celery, carrot, and prune juice; lettuce and tomato juice.
Nervous Disorders	Radish and prune juice and rice polishings.
Neuralgia, neuritis	Cucumber, endive, and pineapple juice; cucumber, endive, and goat's whey.
Overweight, obesity	Beet greens, parsley, and celery juice.
Perspiration	Celery and prune juice; cucumber and pineapple juice.
Rheumatism	Cucumber, endive, and goat's whey.
Rickets	Dandelion and orange juice.
Sinus	Sip lemon juice with a little horseradish; sip mixture of cayenne powder in a cup of water.
Teeth	Beet greens, parsley, and celery juice with green kale.

Disorder	Health Cocktail
Thyroid	Clam juice with celery juice.
Vitality	One T. apple concentrate, one T. almond nut butter, and 1 cup celery juice.
Weight (reducing)	Parsley, grape juice, and pineapple juice.
Youth (retaining)	$2/3$ cup oat straw tea, $1/3$ cup celery, prune, or fig juice, with $1/4$ cup powdered Nova Scotia dulse to each cup. Cucumber, radish, pepper ($1/3$ cup each); $2/3$ cup concord grape juice and $1/3$ cup pineapple juice with one egg yolk.

PROCESSED AND CANNED FOODS

You should recognize that when we boil our foods, preserve them, or process them, we are taking away from their natural food value. They are no longer foods.

Those who depend on "empty" foods, such as some canned foods, should be aware that these products were never intended to provide optimum nutrition. Quite the contrary, they were developed as an emergency stopgap to stave off death by starvation. Food canning was nineteenth century French scientists' response to Napoleon Bonaparte's order to find a way for French armies to carry food. When enemy armies burned fields and food as they retreated, the French soldiers faced death by starvation. Canned foods helped them eat when no "live" food could be found. From this example we see that canned foods can keep death at arms' length. But most people should be able to expect more than this from the food they eat. Still, simply staying away from canned and processed foods in themselves is not enough to ensure disease-free—or even healthful—living. There are many other components to good nutrition, as you will learn.

WARM BODIES AND COLD BODIES

We must understand what effects even the best foods will have on our bodies. This is because even the best foods can be used incorrectly. For instance, we find that many of the fats that people use today are vegetable fats, and such people have cold hands, cold feet, and a cold stomach. By putting hot foods in the stomach, you are going to make a weak stomach muscle structure. You are not going to be able to secrete hydrochloric acid the way you should. You find out that the stomach tissue is not going to respond. The same thing happens when you use cold foods. You are going to constrict the mucous lining of the bowel and you are not going to have the digestive juices there to take care of your foods.

So you see, it's not just using foods that counts—it is also the administration of these foods that is important.

For example, many people do not realize that honey is almost a perfect food in its natural form. But, if you heat honey and feed it to bees, it actually kills the bees. Heating honey reduces its food value so much that it isn't even fit for bees to use. We have to realize that heating natural foods destroys enzymes, vitamins, and minerals. You have to watch what you do with foods.

Many raw foods contain natural enzymes; they can help you to digest your foods. When the digestion is poor, many people use papaya, which is actually a digestant. You find that there are many other foods that you can add to your diet which will help you digest foods.

DIFFERENT FOODS, DIFFERENT FUNCTIONS

Each food is earmarked to do a job for you, creating energy, strength, heat, repairing the cell structure, and so forth. For a detailed analysis of the nutritional functions of more than 100 foods, you may refer to the *Food Remedy Troubleshooting Chart* at the end of this chapter.

Some foods are blood builders. The herb, nettle, is a wonderful blood builder. (All herbs have their effects upon the body and must be considered.) Beets are great for the blood as well. Other blood boosters are green vegetables, black cherries, bee pollen, and sun chlorella. All of these are fine blood builders.

Some foods are brain foods. We help nourish the brain with phosphorus-rich foods, and the same foods that nourish the brain will nourish the nerves and glands. We can use plant phosphorus to build the bones, but the brain must have the higher-evolved animal phosphorus found in egg yolk or codfish roe.

Some foods are muscle foods. We need proteins for the muscle structure in the body. We need the fats that are higher in calories, which give us heat in the body.

Some foods are energy foods. The carbohydrate foods are

sweet in taste, but we do have to chew them. If you don't chew them thoroughly, you won't digest them properly for absorption in the body. You secrete a whole quart of saliva (which contains the starch-digesting enzyme pytalin) in one day from the parotid, sublingual, and submandibular glands, and it comes from a lot of chewing. You chew plenty when you eat raw foods. Conversely, you simply swallow cooked foods or liquids. Raw foods, however, you must chew to moisten them, grind them, and ensure they can be swallowed. In chewing raw foods, we help strengthen the jaw with exercise as well as supply our digestive systems with much-needed roughage that helps the bowel tone itself.

FOOD PRODUCES VITALITY

Do you want vitality? It comes from your foods. And those people who have a lack of immunity—what is it that they actually lack? They lack strength and power: these come from your foods. Vitality foods are foods that give you strength.

Foods can provide heat for the body. People who live in the cold northern climates have to have more heat in the body, and you find out that red foods give them that heat. You find out that cayenne pepper is a great stimulant for the circulation in the body and it helps to bring circulation to the extremities.

The way the body uses fat depends, in large part, on temperature.

Fats each have their own melting points. Vegetable fats sometimes need a good warm body. A cold body cannot take these cold vegetable oils and handle them well. Similarly, a cold body has difficulty digesting cold proteins like beans, lentils, and garbanzos. It takes a warm body to digest them. A cold body can't break down those proteins. So you see, these foods should be used only by people with strong constitutions and warm bodies.

PROBLEMS WITH PROTEINS

Animal proteins putrefy very quickly in the intestinal tract, and that is why we should be careful with meats. Most people don't use enough high-fiber foods to maintain proper intestinal transit time, so we find that meat is one of the most putrefactive foods, even though it is high in protein and relatively easy to digest. It would not be a problem for those with active, normal bowel transit times, but most people in this country are constipated, with underactive bowels. Fermentation takes place in an underactive bowel and favors the multiplication of undesirable toxic bacteria, while the friendly acidophilus bacteria may be almost destroyed. In an underactive bowel, toxic protein byproducts may find their way into the bloodstream, where they cause a great deal of trouble.

We have many kinds of proteins. We have a lipoprotein, a fatty protein. We have a nucleic protein and we have a peptone protein. We have a keratin protein, and these are things that we have to think about. This keratin builds the hair and nails. It is almost an insoluble thing, and we find here that it works along with silicon. We also have a neuro-protein which builds up the nerve cells, and the nucleus of each cell in our body is built with nucleic protein.

Generally, man needs five ounces of nitrates for the muscle structure in the body. He needs twenty ounces of carbonates from starches for the heart, for the heat, for the muscle structure in the body. He needs 3 percent phosphates in the body. He needs one gallon of liquid daily, and then he needs bulk—cellulose fiber that is indigestible but which helps absorb water, adds bulk to the stool, and improves bowel performance.

Fiber is found many times in the seeds, whole grains, fruits, and vegetables. We need this for bowel activity. The bowel develops tone when it has something to work against. It cannot work against slushy, refined foods. It has to have the brans, the bulk, the fiber. This is how strength is developed in the bowel.

FOODS, MOODS, AND FEELINGS

Every organ in the body is different, and every organ, being different, has a different job to perform. We find that the kidneys don't do the work the liver does. The kidneys don't do the work that the lungs do. They have a different chemical makeup. Each organ has a different chemical makeup and each does its job according to the chemical structure that determines and specifies its function. All cells in the vital organs are built of chemicals brought to them by the bloodstream.

The brain centers are different amongst themselves, and are different in each individual person, as well. People who tend to be destructive shouldn't use meat. Meat adds to the destructiveness of a person.

Those who have a lot of affection for people generally turn to sweets; that is why men usually bring candy to their girlfriends or wives. But sweetness is also found in honey, in pies, in cakes. The affection center feeds on sweets and, I might also tell those who are overweight, if you haven't the affection you need, you turn to sweets. It is a satisfaction that comes from the sweets that is missing when our lives are not balanced. And you find out it's like the chicken and the egg story. If you don't have affection, you go to sweets, which causes more trouble with the initial problem. Now if we can develop enough discipline, the problem can be corrected, and the excess weight can be lost.

If you must have sweets, honey is wonderful. It retains the life principles but it is very, very concentrated. Honey is six times more concentrated than sugar, and you have to be careful using it. Dilute it when you use it. Don't cook with it, or the live enzymes will be destroyed, and the honey will be chemically changed. When the cereal is cooked, put the honey on it afterwards. Add it to all foods afterwards, and this is the proper way to use honey.

Temper develops when we use a lot of sauces. Passions begin to rise when we use alcohol. It is the overuse of these substances that we have to think about. Milk is a wonderful food if it is used properly, but most people do not use milk properly. Many overuse milk, since the average American diet is 25 percent milk and milk products, far too much for the body's good. Also, when milk is pasteurized and homogenized, its natural state is altered.

FOOD PREPARATION

Concerning preparation of foods, we have much that we can discuss. I would like to mention here that as far as foods are concerned, dried fruits are six times stronger in their sugar concentration than when the fruit is fresh. The body wasn't built to take these dried fruits. They should be revived by soaking in boiling water. The best way of doing that is to start them in cold water and bring them to a boil. During this time, the boiling destroys germ life and insect eggs—and don't think that dried fruit doesn't carry germs or insect eggs, because it does.

Doctors today are having a lot of trouble with amoebae and all kinds of microorganisms that are found in our foods. So it would be well to revive all of our dried fruits in boiling water.

I mentioned previously about not using hot foods because they weaken the mucous membranes of the stomach and produce spongy gums. I don't mean "spicy" hot, but I mean "hot" hot. Are you beginning to see that all kinds of foods have different effects on the body and mind? You can use your foods as medicines if you begin to see the many things that foods can do.

We also find that cold foods and cold drinks, placed into a warm stomach, draw the heat from the stomach tissue. Cold foods can get you into trouble. Iced drinks and ice cream, as cold as they are, should not be swallowed and left to lie in the stomach to absorb heat. Cold foods and drinks should be warmed in the mouth before they reach the stomach.

When you take too many cold foods, they interfere with the nerve supply from the cerebellum of the brain, and you will find that it will interfere with your getting all the good you should from your foods.

THE TWELVE BODY SYSTEMS

Twelve systems are responsible for carrying on the life process in the human body. These are the skeletal, muscular, respiratory, endocrine, digestive, reproductive, integumentary, lymphatic, excretory, circulatory, nervous, and urinary systems.

Each of these systems has specific functions. Likewise, each system interacts with all other systems to achieve and maintain equilibrium in the whole body. Therefore, if there is a deficiency in one system—for instance, not enough calcium supplied through nutrition (the digestive system)—the body will begin to tap its stores—in this case, stores of calcium from the bones (the skeletal system)—to compensate. This example illustrates the principle of balance on which the human body is based: whatever neglect is shown to one system, the other system will compensate for that neglect.

This law of balance can work in a positive way, as well. Whatever we do to help one system will benefit other systems. For example, if we eat correctly (digestive system), not only will digestion and elimination (excretory system) function more smoothly, but nutrients will be carried in proper balance and generous proportion through the blood (circulatory system) to the glands (endocrine system), thereby aiding all remaining systems to function at their peak.

Through the course of my studies, and as a direct result of personal experience with patients during my 55 years as a clinical nutritionist, I have concluded that the foods, drinks, and herbs listed in the chart beginning on page 60 serve best as natural strengtheners for the specific systems under which they are listed.

Substances most helpful to a particular system are set in boldface. The law of balance notwithstanding, it must be stressed that no one "magic bullet" for lifetime health exists, though many foods do produce remarkable short-term benefits. Rather, good sense and proper nutrition of the whole body over the long-term will produce a state of wellness that safeguards against imbalance in one or all systems that leads to what we call disease.

WATCH WHAT YOU BUY AND EAT

Exercise and work help us utilize our foods more efficiently. Think about that. They help give you that staying power, if you eat right. But we've got all kinds of foods today.

Adulterated foods are common in our time. So are processed and chemically-altered foods, foods with chemical additives in them. We may be using refined white flour that has been excoriated, devitaminized, demineralized, and devitalized. Our bodies are supposed to have *natural* foods, and they will work for you and repair body tissues. If you don't feed your body the way it was meant to be fed, you may end up in *dis-ease*. Dis-ease is a form of disharmony, and that's all it is.

I can tell you about some interesting experiments. The late Dr. Sansome, who started the Sansome Clinic, once produced diabetes in lab animals by feeding them certain foods. He produced hemorrhoids in rabbits and guinea hens by feeding them coffee. He found out that the rectum was affected by coffee. Dr. Sansome produced various disease conditions in the bodies of his animals comparable to many of the common complaints doctors often encounter in patients, and material conditions by processing, additives, overcooking, and so forth.

We don't get the purest, most natural foods any more. Think about it. Manufacturers add dye. They add acids. They add moisturizers, texturizers, artificial flavors, and preservatives. They are now spraying fresh produce and some meats and seafoods in markets and restaurant salad bars with sulfites, which have resulted in several deaths. Sulfites are also found in red wine. Pickles have benzoic acid in them. Luncheon meats have nitrates in them. We need to come back to pure, natural, and whole foods.

The Twelve Body Systems

System	Structure	Function	Vitamins
Skeletal System	All bones, cartilage, joints.	Support and protect body, leverage, mineral storage, red blood cell production.	**C, D,** A, B-Complex, B$_2$, B$_6$, B$_{12}$, E, F, Folic Acid, Niacin, Pantothenic Acid, Bioflavinoids.
Muscular System	All muscular tissue.	Facilitate body movement, produce heat, maintain body posture.	**B$_6$, D, E,** A, B-Complex, B$_{12}$, C, Biotin, Choline, Pantothenic Acid.
Respiratory System	Lungs, trachea, bronchi, bronchial tubes, alveoli.	Oxygenate; eliminate Carbon Dioxide; regulate acid-/base balance of body.	A, C, D, B-Complex, B$_1$, B$_2$, B$_6$, B$_{12}$, E, F, Inositol, Choline, Niacin, Folic Acid, Pangamic Acid, Pantothenic Acid, Bioflavinoids.
Endocrine System	Glands: pineal, pituitary, thyroid, parathyroids, thymus, adrenals, pancreas, ovaries, testes.	Regulate body action by secreting hormones through circulatory system to target organs.	**B-Complex,** E, C, Choline, Inositol, Folic Acid, Pantothenic Acid.
Digestive System	Gastrointestinal tract with exception of large colon (part of the Excretory System), salivary glands, liver, gall bladder, pancreas.	Mechanical and chemical breakdown of food for cellular use.	A, C, B-Complex, B$_1$, B$_2$, B$_6$, B$_{12}$, D, E, F, K, Folic Acid, Inositol, Niacin, Pantothenic Acid.
Reproductive System	Ovaries, ova, testes, sperm.	Reproduction of the organism.	**B-Complex, E,** A, B$_2$, B$_6$, C, D, F.

Minerals	Foods	Drinks	Herbs
Fluorine, Calcium, Copper, Iodine, Zinc, Sulfur, Sodium, Silicon, Iron, Potassium, Phosphorus, Magnesium.	**Sesame seed, kale, millet,** celery, barley, okra, almonds, collards, turnip greens, raw goat's milk.	Black mission figs/raw goat's milk; black cherry juice; green kale juice; celery/parsley juice; veal joint broth.	**Comfrey, kale, boneset,** poke root, chicory, juniper berry, arnica flower, elderflower, oat straw, alfalfa, Irish moss.
Calcium, Potassium, Magnesium, Silicon, Nitrogen, Iron, Chlorine.	Olives, rye, lima beans, rice bran, bananas, sprouts, watercress, complementary proteins (grains, legumes), apples.	Potato peeling broth; dried olive tea; nut milk/liquid chlorophyll.	Juniper berry, tansy, rosemary, black willow, horseradish, wild cabbage, kelp, dulse, watercress, horsetail, black walnut.
Calcium, Iron, Silicon, Potassium, Fluorine, Manganese, Copper.	Garlic, onions, leeks, turnips, grapes, pineapple, honey (eucalyptus), green leafy vegetables.	Celery/papaya juice; carrot juice; watercress/apple juice/1/4 tsp. cream of tartar; rose hip tea; goat milk whey.	**Mullein, elderflower, peppermint, yarrow,** lobelia, comfrey, cayenne, marshmallow, sage, coltsfoot.
Iodine, Silicon, Phosphorus, Calcium, Chlorine, Magnesium, Sodium, Potassium, Sulfur, Iron, Manganese.	Sea vegetables, kelp, dulse, Swiss chard, turnip greens, egg yolks, wheat germ, cod roe, lecithin, sesame seed butter, seeds and nuts, raw goat milk, RNA/DNA.	Pineapple juice/egg yolk/wheat germ/dulse; black cherry concentrate/egg yolk/chlorophyll.	**Kelp, dulse, ginseng, dong quai, licorice,** echinacea, golden seal, dandelion.
Sodium, Chlorine, Magnesium, Potassium, Iron, Sulfur, Copper, Silicon, Zinc, Iodine.	Papaya, liquid chlorophyll, spinach, sun-dried olives, Swiss chard, celery, kale, beet greens, whey, shredded beet, watercress, yogurt, kefir.	Parsley juice; papaya juice; chlorophyll/carrot juice, potato peeling broth; whey drinks.	**Papaya, alfalfa, aloe vera, peppermint,** slippery elm, cayenne, burdock, comfrey, ginger, fennel, anise.
Zinc, Calcium, Iodine, Phosphorus, Iron, Sodium, Chlorine, Potassium, Fluorine, Silicon.	Sesame seeds, pumpkin seeds, seed and nut butters, cod roe, lecithin, egg yolk, raw goat's milk.	Black cherry concentrate/chlorophyll/egg yolk; pineapple juice/egg yolk/wheat germ/dulse; 3/4 cup carrot juice/1/4 cup coconut milk/tbsp. wheat germ oil/tsp. rice polishings.	**Black cohosh, licorice, dong quai, ginseng, blessed thistle,** blue cohosh, uva ursi, raspberry, squaw vine, chickweed, saw palmetto, false unicorn.

The Twelve Body Systems—*Continued*

System	Structure	Function	Vitamins
Integumentary System	Skin, hair, nails, oil and sweat glands.	Regulate body temperature; eliminate waste; temperature, pressure, and pain receptor.	**Pantothenic Acid, PABA, D,** A, B-Complex, B_2, B_6, B_{12}, B_1, C, E, F, K, Biotin, Choline, Folic Acid, Niacin, Bioflavinoids.
Lymphatic System	Spleen, thymus, appendix, tonsils, lymph nodes, lymph vessels and fluid.	Filter blood; produce white blood cells; protect against disease; return protein to cardiovascular system.	**A, C, Choline,** B-Complex, B_1, B_2, B_6, Biotin, Pantothenic Acid, Folic Acid.
Excretory System	Large colon.	Complete nutrient absorption; manufacture certain vitamins; form and eliminate feces.	**A, F, Choline,** B-Complex, B_1, B_2, B_6, B_{12}, C, E, Inositol, Niacin, Folic Acid, Pantothenic Acid.
Circulatory System	Heart, blood vessels, blood.	Distribute oxygen and nutrients to cells; transport Carbon Dioxide and wastes from cells; acid/base balance; regulate body temperature; form blood clots.	**B-Complex, B_6, Niacin,** B_{12}, C, E, Bioflavinoids, Choline, Folic Acid, Inositol, Pangamic Acid.
Nervous System	Brain; spinal cord; nerves.	Regulate body function through nerve impulses; sensory perception and motor response.	**B-Complex,** A, B_1, B_2, B_6, B_{12}, B_{13}, C, D, E, F, Choline, Folic Acid, Inositol, Niacin, Pantothenic Acid, Pangamic Acid.
Urinary System	Kidneys, bladder, ureters, urethra.	Eliminate liquid waste; regulate chemical composition of blood; fluid/electrolyte balance; acid/base balance.	A, B-Complex, B_2, B_6, C, D, E, Choline, Pantothenic Acid.

Minerals	Foods	Drinks	Herbs
Silicon, Calcium, Fluorine, Iron, Phosphorus, Potassium, Sodium, Sulfur, Iodine, Copper, Manganese, Zinc, Magnesium.	Raw goat's milk, black bass, rye, avocados, sea vegetables, whey, apple, cucumbers, millet, rice polishings, rice bran and concentrate, sprouts.	Carrot/celery/lemon juice; cucumber/endive/pineapple juice.	**Oat straw, shavegrass, horsetail, comfrey,** aloe vera, burdock.
Potassium, Chlorine, Sodium.	Green leafy vegetables, watercress, okra, apples, celery.	Potato peeling broth; celery juice; blue violet tea; parsley juice; carrot juice; apple juice.	**Blue violet tea (leaves), chaparral, burdock, echinacea,** blue flag, poke root, golden seal, cayenne, mullein, black walnut.
Magnesium, Potassium, Sodium, Sulfur, Calcium, Chlorine, Iron, Phosphorus.	All squash, flaxseed, green and yellow vegetables, yogurt, kefir, alfalfa tablets, acidophilus, bran, clabbered milk, grapes, whey, psyllium seed, berries, sprouts, yellow cornmeal.	Chlorophyll/carrot juice/coconut juice; celery/parsley/spinach/carrot juice; flaxseed tea; black cherry juice.	**Psyllium seed, aloe vera, cayenne, black walnut,** flaxseed, comfrey, slippery elm, cascara sagrada, senna, barberry, golden seal.
Calcium, Iron, Silicon, Cobalt, Copper, Magnesium, Iodine, Phosphorus, Zinc, Potassium, Manganese, Nitrogen, Fluorine, Sulfur.	Brewer's yeast, garlic, wheat germ, liquid chlorophyll, alfalfa sprouts, buckwheat, sun-dried olives, watercress, rice polishings.	Blackberry/parsley juice; black fig juice; watercress/parsley/grape juice; hawthorne berry tea.	**Hawthorne berry, cayenne, ginger, garlic,** poke root, sassafras, burdock, chapparal, echinacea, red clover, oat straw.
Calcium, Phosphorus, Manganese, Sulfur, Iodine, Iron, Magnesium, Potassium, Fluorine, Zinc, Silicon.	Egg yolk, kale, celery, fish, raw goat's milk, veal joint broth, cod roe, rice polishings, brewer's yeast, nutritional yeast, tryptophan.	Celery/carrot/prune juice; prune juice/rice polishings; raw goat's milk/tsp. sesame, sunflower, or almond butter/tsp. honey/sliver of avocado; black cherry juice/egg yolk.	**Valerian, hops, skullcap, lobelia, lady's slipper.**
Calcium, Potassium, Manganese, Silicon, Iron, Chlorine, Magnesium.	Watermelon (include seeds), pomegranate, apples, asparagus, liquid chlorophyll, parsley, green leafy vegetables.	Celery/pomegranate juice; black currant juice; juniper berry tea; beet/grape juice; celery/parsley/asparagus juice; pomegranate juice/goat whey.	**Juniper berry, uva ursi, parsley,** golden seal, slippery elm, elderflower, ginger, dandelion, marshmallow.

GROWING ENVIRONMENT AND NUTRITION

It has been estimated that about 12 percent of America's food comes from California, which probably has the most over-chemicalized soil in the country. It has also been reported that over two-thirds of California's water supply is polluted. What kind of food can be produced from such a state of affairs? The answer seems obvious.

Let me quote a paragraph from a recent issue of *Solstice*, July/August, 1987, Vol. II, No. 2, which carries a subhead entitled: *Perspectives on Health and the Environment*:

> . . . when we read the FDA's familiar 'Composition of Foods' chart to find out how much iron is in our sesame seeds or how much calcium is in our greens, these figures are completely irrelevant to today's food. They are derived from the 'Firman Bear' report of 1963. The *average* nutritional values found in 1963 were already well below the *lowest* values in the previous study, from 1948. Imagine how much lower they've gone in the nearly quarter-century since 1963! At this point our foods may be virtually 'empty.' We need to create a farming industry that produces food with some content.

HARMFUL FOODS

There are such things as harmful foods produced by our farming industry and the giant food processors. All salted foods, for instance, whether salted fish, salted crackers, salted nuts, or salted seeds, are bad foods.

There are constipating foods. As I have mentioned, blackberries are constipating. Starchy foods are constipating. Beans, coffee, cheese, and chocolate are all constipating in the body, and there are many, many more.

Foods can be our medicine, but not unless they are natural foods. You can change a person's body, but you can only change it for the better by teaching him or her to use the natural foods. If I put a ripe apricot in one hand and a poison pickle in the other hand, which would you choose? You have to decide. You know your body needs good foods. We use disease-producing foods, whiskeys, and spoiled foods that collect germs. Burned foods, smoked foods, fermented foods, rancid foods, and most kinds of vinegars are not good for us. There is a good vinegar which you can get, an apple cider vinegar aged in oak barrels, that comes from New England.

Candy can produce acids in the body. Carbohydrates can even produce an alcoholic condition in your body. William Jennings Bryan, one of our great orators, at one time went around the country telling people how bad whiskey and the drinking of alcohol was, and he died of an alcoholic stomach. The day he died he had eaten thirteen hot cakes with syrup for his breakfast. It produced an alcoholic ''still'' in his stomach, and he died from alcohol poisoning! It's hard to believe that these things can actually happen.

Then we have eliminating foods. We know that prunes are eliminating. They have a certain acid in them that makes prunes an eliminating food. But prunes also have the highest food content in nerve salts. Prunes are a very good food.

THE THREE WORST FOODS

I would like to mention here the three foods that every wise nutritionist leaves out of a diet:

1. *Rhubarb*. It is high in oxalic acid and it adds to joint disturbances in the body, especially when it is heated.
2. *Cranberries*. These are very high in oxalic acid, but most of us don't eat them raw; we have to have them cooked. We allow cranberries on Thanksgiving, and that's enough.
3. *Green Plums*. This is another one of the negative foods that should be avoided at all costs.

Now I would like to tell you that green plums, rhubarb, and cranberries are three foods that the birds won't eat. When the American Indian wanted to know what he should eat, he would watch the birds. If the birds would eat it, he knew that he could eat it. Isn't that something to think about? Since the birds won't eat them, perhaps they should be left out of our diet.

FOODS AND THEIR FUNCTIONS

Beets are quite eliminating. They are laxative in nature. Apples and radishes are eliminating foods. We have a laxative food in flaxseed. Many times I tell my patients to grind it and put one tablespoon in with their cereal. It is also a wonderful laxative for children. Flaxseed will keep that bowel on the move.

Barley soup, bacon, and sugar are all fattening foods. Pork and other fat meats are fattening. Even buckwheat is a fattening food. So, if you want to reduce, you are going to have to cut down on these foods.

I recommend avoiding one food—wheat—simply because it is so high in gluten, and because most people use too much of it. The average American diet is 29 percent wheat—mostly refined wheat products, low in food value. Gluten can damage the intestinal villi, causing celiac disease, which causes malabsorption (and sometimes bleeding) in the bowel, and results in calcium deficiency. The only known cure is to avoid gluten-containing products, including all wheat products.

The sulfur foods, such as cabbage, cauliflower, Brussels sprouts, broccoli, radishes, and eggs, are gas-producing foods. Burned foods and greasy foods produce a lot of gas.

Then we've got solvent foods. Do we need them these days? We certainly do, with all the cholesterol problems and hardening of the arteries. Some good solvent foods are lecithin, fish oils, and chlorophyll.

Berries are also wonderful laxative foods. Nettle broth is laxative. Cherries are very high in iron, but they are laxative to the body.

We have summer foods and winter foods, and these come along with the season. Strawberries are summer food. All squashes are summer foods. Apricots are a wonderful laxative food, high in copper, which we need to build a good red blood count in the body.

There are a few foods, such as asparagus, that quickly react with the body. For instance, only thirty seconds after eating asparagus, it can be smelled in the urine. It has that strong an effect within the body. Likewise, while I was studying in Germany, I witnessed a demonstration in which a garlic clove was put in a person's shoe; in less than thirty seconds you could smell the garlic on that person's breath. This is passing through the body pretty fast, isn't it?

THE WONDERS OF SOME WONDERFUL FOODS

Those who have broken capillaries should know about rutin, which comes from buckwheat in a concentrated form. The walls of the tiny blood vessels are very fragile. Imagine using foods now to take care of that. Barberries contain barberic acid, which tones the capillaries, as well. Silicon is most important for the repair of varicose veins. Barley is not only a good heating food for the winter time, but also very wonderful for the nerves and muscle structure. Beans give staying power for the muscle worker.

Veal has almost as much protein as beef has, but not as much uric acid. Blackberries are especially good for the cerebellum. Bran tea is where you get your silicon, which is needed by the hair and nails. Buckwheat is a winter food, a heat producer.

Buttermilk is a wonderful laxative for the intestinal tract. You can even go so far as to use buttermilk internally as an enema. There is nothing better for an acid bowel and itching rectum that has an acid rash with it than buttermilk enemas. Buttermilk will neutralize all of that acid, which can be so irritating in the rectal area.

JUICES AND TONICS

To mix the juice of carrots with whey is the most wonderful combination for the intestinal tract.

Cauliflower is easier to digest than cabbage, and provides enough sulfur for the body.

Celery can be a brain tonic, and the sodium in it will neutralize acids in the body. It is one of our highest sodium foods.

Most people know that cheese is binding. But why is it binding? We know that when the cheese is made, whey is a byproduct. The whey is removed from it. Whey is laxative, and this is one reason why we use Whex—a concentrated dried goat whey. Whex is possibly the highest source of organic sodium. It can be used to build up the acidophilus bacteria in the body. It may not act as a laxative on the first day, but you will find in time that it will balance the different flora we have in the bowel.

Berries are also excellent for that bowel. One of the things that berries will do most for you is keep the blood from clogging. Isn't that something to think about?

Barley and green kale soup is one of the greatest calcium builders in the body.

For rheumatism, one of the best things is cherry juice and chlorophyll together. That combination is also great for the liver. The liver is an iron organ. If you use chlorophyll as found in greens, this will straighten out the liver.

Coconut is 70 percent fat, so you have to be careful not to take too much. You have to have a balance in everything you do.

Cream is excellent for the stomach and can be added to carrot juice. It can be added to broth, but it is cholesterol-forming the moment you bring it to a boil. So, add the cream after the soup is cooked and brought to the table. This is the way we should have it. Cream is a wonderful thing for the stomach, the lungs, and the sexual system.

Cucumbers are a summer food. Have you heard the old saying about being "cool as a cucumber"? Why is it cooling? It is high in sodium. There is truth in that saying.

Some people have said in the past that even fish is a brain food, and this is true. In fish we find phosphorus for the brain, and this is why we consider it as a brain food.

AN EXTREME CASE AND A HAPPY ENDING

Many years ago I was called to see a little lady who was paralyzed and in bed. She weighed only about eighty pounds. She couldn't speak or move her body. She was dying from cancer of the breast. Now this was many years ago before the laws were passed about not treating cancer. I was brought to this lady by a minister of the Science of Mind Church in Pasadena, California.

I was told that the lady had only three weeks to live. Is this enough time to tackle such a condition with foods? How much can you do with just foods? Well, what was I going to do? If a person has a spark of life in them, then you must still work with them, because more hope is found in the spirit than in the body itself. The worst situation a doctor can face is to take care of a person who knows that they are going to die.

The very first thing I did was to take the covers off of her feet and look at them. I made a face, because her toes were getting blue. You know, we die from our feet up. Then I put the covers back.

I gave her some Whex (dried goat whey) dissolved in warm broth and started giving her one tablespoonful about every hour. Whex is very high in potassium, sodium, phosphorus, and calcium.

For the last four or five days she had been unable to eat because she couldn't get anything down. But, working with this little lady, I finally got her to take the Whex and broth. I sat with her through the night, feeding her a tablespoon every hour.

The next morning she had new energy. She was better than she had been the day before. I continued working with her, and day by day she gradually improved. In about three months she regained almost twenty pounds.

When she went back to her own doctor, he couldn't believe

what happened. She said, "I want you to give me a checkup." The doctor examined her thoroughly and said, "I don't see any sign of cancer in that breast. I must have made a mistake in my diagnosis." Many people don't seem to understand that Nature heals, but sometimes it needs a helping hand.

I want to tell you what made the difference. Sometimes doing nothing at all is the best remedy. Hippocrates knew this. You've heard the old saw, I'm sure: "Treat a disease and you will get well in two weeks; don't treat it and you will get well in fourteen days."

Of course, I did give this lady a little broth and Whex to bring up her strength and energy. After six months' time, I got a card saying she was out mowing the lawn. She was supposed to have been dead in three weeks when I first saw her!

Later I heard her explain to a group of people what had happened. "After three doctors had given me up to die," she said, "Dr. Jensen came to see me with the minister of my church. I saw the look in his face when he lifted up the covers at the foot of my bed and saw my toes were blue from lack of circulation. But he stayed with me. And I made up my mind I was going to show those doctors. I decided to show them I wasn't going to die! I just depended on Dr. Jensen, and here I am today!"

Isn't that a wonderful testimony? You see, having the right spirit is one of the keys to leaving a disease in the past. But we must keep in mind that when a patient is weak, they are unable to make wise decisions about what to do. As Hippocrates said, "Physicians should take the lead in showing what is needed to restore health. If left under orders, a patient will not go far astray."

OTHER WONDER FOODS

I would like to mention eggs. The egg has the right nutrients for the brain, nerves, and glands. I would also like to mention dates, a heat-producing food, and garlic, an antiseptic food. No harmful microorganism can live in the presence of garlic. High in sulfur, garlic makes the brain more active. That is also why it is good for the sexual system. I used to grow grapes at my sanitarium years ago. In Europe people go to the grape-growing regions, rent a hotel room, and take what they call the "grape cure." They eat nothing but grapes for days and days. They begin to eliminate, to get rid of accumulated catarrh and toxic material. We have to cleanse the body of old toxic material before new tissue can come in place of the old.

We had a lady at the sanitarium, a nurse from Salt Lake City, who had a cancer of the breast. She came to the sanitarium and went thirty days on grapes, and all signs of the cancer disappeared. To see this happen is an unbelievable thing. By profession she was a registered nurse in a hospital.

I can tell you of many cases where I could see the correction come, even in the worst cases, with foods alone. But you know, taking care of extreme conditions with foods is very difficult. I don't like to say that we're going on carrot juice and whey exclusively. But I want you to know that foods have brought many people back after their doctors had given up on them.

The best grapes are those that have the seeds in them. On the outside of the seed there is a tartaric acid, a cream of tartar. That cream of tartar is the greatest thing to cut the mucus and catarrh so they can be eliminated from the body. Catarrh can be taken care of a good deal by just adding a little cream of tartar to the diet. Like whey, it isn't a whole food, but its activity is very effective.

Gruels are necessary in many intestinal problems, and a rice gruel is often used in China. Barley gruels are also soothing and healing. We are going to discuss some of these foods later.

Flaxseed tea is one of the most soothing teas for the intestinal tract. Take a cup of flaxseed tea and put one tablespoon of liquid chlorophyll in it. The chlorophyll cleanses the bowel and feeds the lactobacillus acidophilus.

Most honey is digested and assimilated very rapidly, which makes it unfit for diabetics. But Tupelo honey from the South is digested much more slowly, and is said to be acceptable even to diabetics, to a certain extent.

Horseradish is wonderful to break up a sinus catarrh. Put a little powdered horseradish on the tip of your tongue, breathe in, and the whole top of your head will come off! My patients often do something similar with sage. They would pick fresh sage right off the bushes beside the road, rub it in their hands and breathe deeply with their hands cupped over the nose. This opens up the sinuses very nicely. Or, you can use bay leaves.

VARIETY IS BEST

Now we know that the legumes are high in potassium and phosphorus but poor in calcium and sodium, and this is one reason why we have to have a variety of other foods. Otherwise you won't have a good balance for the whole body. In other words, the whole body has to be fed. You are made from the "dust of the earth" as assimilated into plants, not the dust from your backyard. Do you understand? And that means Washington apples, Pennsylvania plums, tomatoes from Texas, and celery from the Salt Lake valley— a variety of foods from a variety of places.

WHAT WE NEED TO KNOW ABOUT MEATS

Those who eat meat must be very careful these days because of the chemicals and cancer-causing hormones (such as diethyl stilbestrol—DES) that are used in livestock feed. Always get the young meat and use more chicken and fish than red meat. If you have access to wild game—venison, wild turkeys, ducks, and so forth— then use that. Wild game has a much more varied diet and is more nutritious. Wild meat is very high in potassium and alkalinizes our body beautifully.

Always select lean meat and avoid fatty meats. By the time fatty meats have been cooked, the lecithin is destroyed, but the cholesterol remains. Since lecithin is needed to keep cholesterol in liquid form in the bloodstream, the absence of lecithin in fatty cooked meat contributes to fat deposits on the artery walls.

Red meat is stimulating, sometimes too stimulating. Scientists recommend less red meat and more poultry and fish in the diet. Fish has oils that help reduce the risk of heart disease. The Japanese, who eat a great deal of fish, have a much lower rate of heart disease than we do in the United States. Some have said that the Japanese are more naturally resistant to heart disease, but that is not necessarily true. Japanese-Americans have the same high rate of heart disease as other ethnic groups in America.

When you select fish, pick fish with fins, scales, and white meat, such as trout, sea bass, halibut, red snapper, and black bass. Salmon, which may be pink or red, is also a highly nutritious fish to eat.

Meat, poultry, and fish should never be fried, but always broiled, roasted, or baked. All fried foods should be strictly avoided in order to reduce risk of heart disease and cancer.

THE ADVANTAGE OF NATURE'S WONDERFUL FOOD VARIETY

Olives are the highest potassium foods. If they are so high in potassium, then we ought to know how to use them. Here's how:

Get some sun-dried olives, put them into water and revive them. Then put them in olive oil again, and you can get that ''natural'' olive back without that strong acid which was once there. Potassium neutralizes acidity in the body, especially the muscle tissue. Potassium is a heart element.

You can take ten olives, steep them in a teapot of boiling hot water, and you'll have the highest potassium tea that you can possibly make. This is how we can help keep that heart well. Potassium tea from olives is a lovely natural remedy, and you are going to learn to use foods as your remedy.

Foods used moderately and wisely have no undesirable side effects in the body. Parsnips are wonderful for bowel activity. Pears are a summer food high in iron. With potatoes, we should be a little careful. It is a hard starch for many adults to take. The sweet potato is very constipating in its nature. But it is better than the white potato. Rice is easier to digest. We should use more whole grains and a little less potatoes.

Consider that raisins have less acid in them than grapes. The acid has been dried out of them. We find that raisins are very high in sugar, so be careful of them. Always brush your teeth after having raisins to reduce cavities.

Salmon is one of the highest foods in the long life factor—RNA, or ribonucleic acid.

Watercress is a great blood purifier. Whey is a great food in its benefits for the intestinal tract.

Bananas are high in potassium. I don't know if you realize that if you examined the center of a banana you would find the seeds are almost like glass. They are tiny, but if you take them and run them between your fingers, you can actually make your fingers bleed. I mean, this is a hard seed that is in the banana. It can be very harsh to the intestines. But is is a very wonderful food for getting the potassium into your body.

We have 400 varieties of avocados today. I would like to mention a note of caution. Use the fats and oils as found in the whole foods. Avoid using any concentrated oils as foods. We are now getting into trouble with olive oil, safflower oil, sunflower oil, and other concentrated oils. It takes ten avocados to make one teaspoon of avocado oil, but one teaspoon of pure, concentrated avocado oil may harm the liver, while if you took ten avocados the way that you were supposed to eat them, the liver would not be harmed.

If you take a green plum, such as the Santa Clara plum, and dry it, you will find that this dried fruit is better for you. Some plums are very high in acid, and I advise avoiding them.

NUTS AND SEEDS

Nuts are high in natural oils and are heat-producing. The almond is the king of all nuts, but many kinds of nuts are nutritious. Hard-shelled nuts are best, and least likely to be rancid.

Here is a way to get out the natural oils. There is oil in almonds, oil in cashew nuts, and oil in coconuts, and these can be used as seasonings in our various foods. Use grated almond, grated coconut, or grated cashew as the seasoning, not the concentrated oil alone. The almonds have the oil in them. Always use the hard-shelled almonds, not soft-shelled nuts, because you will find that the oil of the soft-shelled almond is dried out or rancid. We need the oils and the lecithin that are in raw nuts. Oil and lecithin are the brain, nerve, and gland foods that you have to have, and you can get some of them from hard-shelled nuts.

Cashews come from India. I don't think you can get raw cashews from India because I believe that they have to go through a fumigating process before they can be exported. I'm just a little careful of cashews. However, they are wonderful nuts and quite nourishing.

Then we come to seeds. Sunflower seeds can put weight on you, but sesame seeds will not. We have many kinds of seeds, but I consider the sesame seed the champion of all seeds. You have to be careful of coconut meat these days because it is sweetened. It also contains glycerine that is put in it. This glycerine is a coal tar product, and many of the coal tar products are cancer-producing.

FIGS, DATES, AND DRIED FRUITS

Black Mission figs are among the highest sodium foods. Figs have a mild laxative effect. They do a doubly-helpful job in building up the "friendly" acidophilus bacteria in the bowel. You find, along with raw goat's milk, that figs are high in an organic, natural sodium that you can't buy in a bottle. Remember, you can get your best remedies from your foods, if you know what you are doing.

I believe that dried fruits make wonderful natural sweeteners, but they should be revived by putting them in a pot of water on the stove. When the water comes to a boil, turn off the fire and let them soak overnight.

The greatest sweetening for whole grain cereals is steamed raisins. You may want to cut up dates for sweetening. Dates, in fact, make an attractive after-school snack for children, especially if they are stuffed with nut butter. You can also get a pure maple syrup that is simply wonderful.

BE CAREFUL WITH CITRUS FRUITS

Now I do not approve of too much grapefruit or oranges. First of all, citrus is usually picked green. It is picked about six weeks before it would normally be tree-ripened. For that reason I don't recommend using much of it. You could have it once a week or once every ten days if you use other juices and fruits on the other days. Do you see the importance of variety?

When you go to Hawaii, you stuff your tummy with all the papayas, mangos, and pineapples you can manage, then in a week or two you are breaking out in boils and pimples. These are eliminating foods. They are extreme eliminating foods.

Why does your skin break out when you eat eliminating foods? Because the kidneys can't take care of the excess juices. Did you know that? If you want to take care of the skin, you have to take care of the kidneys, but if you want to take care of the kidneys and help them, then you have to take care of the skin as well.

HOW TO TELL WHEN FRUIT IS RIPE

Here's a way to tell if an avocado is ripe. Put all five fingers on the avocado and squeeze it. If you can feel that the seed inside is separating from the meat, it is ripe. With five fingers you won't bruise it. When the seed within the avocado begins to separate from the inside meat, then you have a ripe avocado.

I also want to bring out that when you are shopping for a ripe pineapple, there's a way to get it to be uniformly ripe all the way through. Often all the ripe juices are at the bottom end, while the top part with the fronds is still green. If you want to get that whole pineapple ripe, set it up on the frond end, and the whole pineapple will get ripe. When the little fronds can be pulled out easily, it is a sign that that pineapple is ripe. Now a ripe pineapple is a hard thing to buy, because they don't pick them ripe enough.

There's a way to tell when red apples are ripe, too. When you are looking for ripe Delicious or Jonathan apples in the store, examine the "flower" end of the apple (the opposite of the stem end). If the apple is red all over, but green in the flower part, you'll find it isn't a ripe apple. The coloring has to be red right through to the flower end.

You should know that 75 percent of the insecticide spray that is used on apples ends up in the core of the apple. So don't eat the core.

MORE ON SWEETENERS

There is a turbinado sugar that is a natural, raw sugar. People don't think about the raw sugars today, but you should get the rawest sugar that you can possibly find. There is a "woolsen" factor that is removed from processed white sugar, and that can cause problems. People who use raw cane sugar never have arthritis or rheumatism. People in the South who use raw sugar cane don't have troubles with rheumatism and arthritis. But when the "woolsen" factor is taken out, we find that processed white sugar users are vulnerable to rheumatism and arthritis.

Of course date sugar is very good as a sweetener. When I was in the Hunza Valley in India, I noticed that apricots were often eaten. There were fresh apricots in summer and dried apricots in the winter time. To use the dried apricots, they revive them in a little water, rubbing the apricots against the side of a wooden bowl to make a kind of purée out of them. When they are ready to drink it, they add a tablespoonful of buckwheat flour as the final touch. And they just love it!

Again, you will recall how I mentioned that dried apricots, raisins, dates, figs, and other dried fruit make such great sweeteners for our whole grain cereals. There are other supplements we can add as well for increased nutritional value. We can, for example, add grated raw sesame seeds, almonds, sunflower seeds, or pumpkin seeds. We can add rice polishings, high in silicon, B-complex vitamins, and natural fiber.

SEEDS AND MEALS

There are four whole grain cereals I recommend as alternatives to wheat. It's nice to use oatmeal occasionally. But, like wheat, it is high in gluten. I consider yellow cornmeal to be a wonderful substitute for wheat. Yellow cornmeal has 4 percent phosphates in it while white cornmeal has only 2 percent phosphates. Phosphates feed the brain, nervous system, and glands. Brown rice and millet are excellent cereals. Millet is the least fattening of the grains.

I might just tell you that flaxseed is one of the highest foods in vitamin F. Flaxseed can be made into a tea. Vitamin F builds our intestinal tract and feeds the villi that the gluten in wheat and oats tends to break down in the intestinal tract.

To make a flaxseed tea, add one tablespoon of flaxseed to one pint of water. Boil it for eight minutes, strain the seeds, and drink it. If you have intestinal trouble and you want to use flaxseed, don't boil it. Just put it in hot water the night before and soak it all night, then strain it the next morning. If you boil it again you will destroy the vitamin F.

TOMATOES AND SQUASH

We have different kinds of tomatoes. The yellow variety is the best to have because it is a non-acid fruit. Tomatoes, like citrus, contain citric acid, and too much is not good for us. So you have to be careful of it. How do you know whether a tomato is ripe or not? When you cut it in half, if the seeds are brown, it is ripe. But if you cut into the tomato and the seeds are green, it isn't ripe. Tomatoes are wonderful sliced or cut up in salads or even canned.

Now the squashes. The zucchini, the yellow crook neck, and all of the other squashes are the greatest things for bowel troubles that I can tell you. They produce absolutely no gas in the intestinal tract, not even the big banana squash. All squashes are low in calories, high in fiber, and easily digested.

SOUP'S ON!

One of the best soups is made with fresh corn. Cut the corn off of the cob and put it in your liquefier. In this way, make a raw corn soup. You can make a raw asparagus soup with a little cream or use a little oat straw tea or an herb tea. You can make a nice soup or broth from that. Some of your best soups can be raw.

There is no reason in the world why you can't have a raw spinach soup. These are just a few of the ideas you can add to your own.

Under the outside half-inch of the potato, 60 percent of the potassium is found. If you boil a potato, cut it in half, and look at the outside edge. You will find a white ring under the skin. This indicates the presence of potassium. This outer part of the potato is useful for making potato peeling broth. This is the broth highest in potassium. Potassium neutralizes the acids in the body, and this includes the lymph stream too.

To make potassium broth, all you do is take two potatoes, simmer in a pint of water for twenty minutes, strain off the peelings, and drink the broth.

A RAINBOW OF ENERGY MEDICINE

The day is coming when we will find out that we are living on the electromagnetic energy of our foods, not exactly the bulk itself. Each food has its own unique electromagnetic contribution to make to the processes which continually ''re-create'' life and health in us day by day. Each food has a unique function. Likewise, the uniqueness of each food is also represented by its color.

I believe that the carrot is the most neutral of all the vegetables we have. Very few persons ever have problems digesting or assimilating carrots or carrot juice. That is because the color and energy of the carrot help make it compatible with the human body. I have seen people get wonderful benefits from carrots.

We have a whole rainbow color spectrum of foods. We have the white cabbage and we have the red cabbage. (I've already mentioned that cauliflower, a sulfur vegetable like cabbage, is easier to digest and less gas-forming than cabbage.) We have purple beets, orange carrots, green lettuce, white grated parsnips, almost a rainbow of vegetables to make salad from. We can use the red bell pepper, yellow (or red) tomatoes, green celery, and others. What do I tell you when you have a salad? I call it a ''rainbow salad.'' You may not have every color in that salad, but in a week's time you can include all the colors representing all the variety of chemical elements we need. Red cabbage contains chemical elements different from those in the white cabbage, but we need them both.

Here, I would like to mention something about the green cabbage for a moment. There is 40 percent more calcium in the outside leaves of cabbage than in the inside leaves. Now we need this calcium, so don't throw away the outside leaves. Use them.

When I select lettuce I always go for the darker green lettuces. Greens are the healers in the body. They bring in chlorophyll,

carotene, and calcium, all important in the healing process. I avoid
the pale green iceberg lettuce, and personally I don't like Romaine
either. It's a little tough for my taste. I use a lot of energy and I have
a sensitive body, so I like to use the Bibb lettuce. I like to use the
red-tinged lettuce, and this type is high in chlorophyll also.

When we talk about salads, we have to talk about sprouts.
Sprouts are very high in silicon, as shown by the shiny, slick surface
of the stem. Keep in mind that sprouts come from seeds that
contain all the elements necessary for life, and when these seeds
first begin to sprout (usually between five and seven days), they are
rich in vitamins, minerals, enzymes, and prostaglandins. You can
put them in the refrigerator and they will stop growing.

One of these days we're going to study darkness. All life comes
out of that blackness and reaches for the light. Light is necessary to
develop the chlorophyll that we find in the greens, in the small
leaves on the tops of sprouts. Sprouts will do more for constipation
than anything I can tell you about. They have the bulk to do it.

Tofu, a soybean protein, is often used in salads these days. I
don't particularly like tofu because it is made with a magnesium
chloride salt. That is inorganic salt. If you could get a tofu that
wasn't made with magnesium chloride, it certainly would be a
wonderful thing. Personally, I can wait.

KUMQUATS AND LOQUATS

The loquat is a little bit larger than the kumquat fruit and it is
wonderful when it comes in at the beginning of the season. The
loquat or the kumquat is very bitter if you just take one bite. But if
you put the whole thing in your mouth and start chewing it, it is
very sweet. The bitterness was in that peeling, but the inside
brought out that sweetness.

Kumquats and loquats have seeds inside, and you should chew
the seeds too, because this is where the vitamin E and the factors
similar to hormones are. I wish I could share more with you about
foods with seeds, because seeds are the glands of plants. The foods
that have seeds are the greatest foods for us. Seedless foods are
out—they've taken the glands out of them. Foods with seeds help
nourish the glands, nerves, and brain.

THE SEEDLESS GENERATION

The reason that we have so much pornography and sex problems today is that we've been using too many seedless foods. Did you know that? When we don't feed the brain and reproductive system with the proper nutrients as found in seeds, a tendency toward sexual imbalance and perversion develops.

Oranges were once fruits as small as berries, full of seeds, but the original orange was corrupted to develop a large, seedless fruit. Now the next generation is depending upon these seeds, but it isn't getting them. We are using seedless foods. It's a sad thing. Seeds have all the chemical elements, nutrients, and nucleic acids to produce new life. Kirlian photography reveals energy emanating from seeds that demonstrates the presence of the life force in them.

We need the electromagnetic energy that we get from our foods. Berries, I think, are among the greatest foods we have, because they are loaded with seeds. The fruit part of the berry is loaded with vitamins and minerals, and the seeds add another dimension of nutritional value.

A final note on berries. When you get your own ranch, let me suggest that you grow what they call ''Golden Raspberries,'' the sweetest and loveliest tasting berries you can imagine. Try them. They are just for you!

All right, we find that almost all berries have a laxative effect, but the effect that we see from the seeds is absolutely wonderful. Watermelon seeds are good. Unfortunately, these days they are even growing seedless watermelons. They are going to take all the gland material out of watermelons. In Texas they have a new way of injecting the watermelon so that the red will go all the way out to the green skin. Most people only eat the pink part of the watermelon, down to the white, unripe part near the outer green skin of the melon. So they are going to fix up watermelons so you can have the red clear out to the outside.

The white in the watermelon is often used to make watermelon pickles. Coming from a Danish family, I used to watch my mother make them. She cut up the white part of the rind and made pickles out of it. The white in the watermelon is one of the highest sodium foods you can possibly get. Can I also tell you that one of the highest sources of chlorophyll is the peeling on the outside of the watermelon? Watermelon is an unusual, valuable food.

You can save watermelon seeds, put them into a blender, grind them up, strain off the hulls, and make a seed milk drink.

Every year Canadian fruit buyers purchase plane loads of Florida watermelons. They bring these melons back to Canada for the seeds, because these seeds are the greatest things for the kidneys that you can put into the body. You can go to health food stores and buy watermelon seeds. People buy them because they are tranquilizers for the body. It is the finest thing for the nervous system. Most people just spit them out and throw them away. It's no wonder that we don't keep the kidneys well, that we don't have enough sodium for the joints. This is because we do not eat the whole food, the whole watermelon, seeds and all.

Select a female watermelon for greater sweetness. How can you tell a female melon? Melons have a flower end and a stem end. If the end is indented, that is a female watermelon. That is the sweetest melon. Those that come to a point at either end or both ends are male melons, and they are not the sweetest melons. That's something to think about—the indentation where the stem is attached.

The gentlemen may be especially interested in this news: the Cavalda Date Company has reported that the date seed is the highest seed in the male hormone. Similarly, seeds from citrus fruit are the highest in the female hormone. Seeds can have either the male or the female hormones.

Every living person has both male and female hormones. If women have too much of the male hormones, it is bad, and if men have too much of the female hormones, this is also bad. There are some women who have grown a mustache. There are some men who have a full head of hair, and some women who are bald-headed. This comes about because of the glandular imbalance in the body. I'm bringing this out because I don't want you to neglect the seeds for your body.

WHAT FRUITS CAN DO FOR US

Strawberries are among the highest fruits in sodium. Like most other fruits, and some vegetables, strawberries are eliminating. Here is something you should know. To stop any elimination process, take a little protein or a starch with it. This will stop the elimination process. Some of us are not building up the body as well as we should. There is such a thing as too much elimination.

Mulberries are unusual berries—unusual in taste, unusual in the quality of the juice. You have to be especially careful using mulberries because, if you get them on your clothes, you'll never get the stain off again. It is a stain that goes on and stays on.

Mulberry juice is the greatest thing in coating the stomach wall when we have a stomach ulcer. It is the greatest thing for taking care of mucous membranes in that stomach wall.

Persimmons are excellent, as well. At my ranch I have a statue of the Chinese god of longevity, who has a persimmon in one hand and the staff of life in the other hand. They claim the greatest fruit in China is the persimmon, and it is considered the "long life" fruit.

The Japanese also have a persimmon, called the Fuji persimmon. It is flat like an apple, and you can cut it in flat pieces. I grow both kinds on my ranch. Persimmons can be used to make a frozen dessert that is a wonderful substitute for ice cream, and much better for you.

I learned from the Japanese how to fix persimmons. When we pick a ripe persimmon it has a little bit of a knob on top. We dip it in hot water, then take off the peeling. It is a fine peeling, but that fine peeling on the outside is all potassium, very bitter. After the skin is off, you can hang them on a string. My wife, Marie, hangs persimmons on a clothes line to dry. By the time they are dry, they have a layer of natural sugar on the outside, a white coating. This is a tremendous and economical natural candy to give to children. A dried persimmon is expensive, about $1.98 in a health food store.

We can also dry strawberries. We have dried mulberries and we even dry rose hips. We want the vitamin C these dry fruits provide. With a little common sense and some hard work, we could give up all this vitamin pill stuff. Do you understand that? I want to emphasize that you can stay well with the direct benefits of the proper foods. If you know your foods, you will recognize which healing properties they have. What will prevent an illness will cure an illness.

MORE FOOD EXPERIMENTS AND EXPERIENCES

I use vitamins and minerals but I only use them over the short term for deficiencies. Over the long term, you must learn to get your vitamins and minerals from foods for the best effect on your health.

In clinical experiments, three medical doctors from New England administered pumpkin seeds to a group of men with cancer of the prostate. After a period of eating pumpkin seeds, the cancer patients' disease actually regressed. These results show that pumpkin seeds in the diet may actually prevent cancer of the prostate.

Swiss nutritionists reported similar success with beet juice. Experiments they have done reveal that beets cause a greater flow of bile from the gall bladder and the liver. In experiments with beet juice on rats and guinea pigs, these nutritionists discovered that the test animals were not prone to cancer when cancer cells were injected into them, if they had eaten a steady diet of beet juice.

GOAT WHEY FOR LONGEVITY

I am certain you have heard of the old men I have visited around the world. Whey is the drink that these old men in Russia, Rumania, and Turkey drink, and that whey is the highest thing in natural sodium.

Whey keeps you young and active, keeps the joints youthful and pliable. I have never found an old man who used whey who had any arthritis or any digestion problems. Some people can't stand the taste of fresh whey, but you can add it to some carrot juice or other juice. If you want to, you can dilute it with green juice for extra iron. Whey doesn't have much iron in it, but we usually don't drink it for the iron. We drink it for the sodium.

USE GELATIN TO INCREASE YOUR CALCIUM INTAKE

Gelatin is one of the best sources of calcium. This is a nice way of taking soluble calcium into the body. Whey is the way to take soluble sodium. If you want a good sodium meal, you take whey with some nice black mission figs and you've got a great sodium combination, wonderful for taking care of the joints.

I started mentioning gelatin. Gelatin is actually 48 percent calcium, and you will find out if you watch dogs that they always eat the ends of the bones where the calcium is located. We need calcium as much or more than they do. If you can prepare a gelatin mold and use a cherry concentrate in it, you don't have to add any sugar. That is a lovely food, especially for rheumatism. Gelatin is what you have to use to build up your body when you get arthritis, rheumatism, osteoporosis, and so forth.

We can thank Hippocrates for much of this knowledge. I am Hippocrates' student because I can make food work like a medicine. And this is what he said: "Food can be your medicine, and medicine can be your food.'' Nutrition is the health art that the future will rediscover. Medical colleges are going to be teaching more courses on nutrition in the days to come. Do you know why? There is so much pressure coming to bear on them from people who want to live correctly that they have to do it. It's only a matter of time.

People are tired of raising children who become doctor bills. It is time now that we start bringing up children who are healthy and well. There is no reason in the world why we should have such extensive health care bills in this country. I can say that 75 percent of them could be wiped out by the proper nutrition. The American Cancer Society says 35 percent of cancers can be controlled. Food has an influence on cancer. If more of our nutritional knowledge were used in taking care of cancer, I'm positive that we could go to 75 percent or higher in controlling cancer.

I believe in foods. I believe in foods as our best medicine. The following Food Remedy Troubleshooting Chart should aid you in your choice of natural medicines. I hope this information can be an awakening for you and I hope again that what I've presented throughout this book has encouraged you to get into this wonderful work.

On the wall at my ranch is a painting of Li Chung Yun, a Chinese sage who lived to the ripe old age of 256 years, teaching for many years at a Chinese university. Arrangements have been made for me to visit that university on my next trip to China. I'm looking forward—with great pleasure—to walking through the university campus where Li Chung Yun once taught. It will be a great honor for me.

Yun, who died in 1930, was also a student of the natural healing art, who relied a good deal on herbs, and on a philosophy of inward calm. Like Hippocrates, Rocine, me, and millions of other people the world over, he understood the tremendous stores of potential energy in foods. I believe that, by combining this knowledge with a well-balanced mental outlook, we can elevate the quality of our own lives and, in so doing, elevate the quality of life on earth as a whole.

Food Remedy Troubleshooting Chart

Food & Type	Predominant Chemical Elements	Best Way Prepared and Served for Digestion	Remedial Measures
Almond Nuts Protein Fat	Manganese Phosphorus	Serve with vegetables or fruits. Almonds, celery, and apple: a complete meal.	Muscle, brain, and nerve food. Best of nuts to use.
Apples Mineral Carbohydrate	Potassium Sodium Magnesium	Wash, eat alone, in salads, or with proteins. Give to children in between meals.	Apple skins used for tea. Fine for kidney and urinary tract.
Apricots Mineral Carbohydrate	Potassium Phosphorus Iron Silicon Copper	Use only fresh or dried (unsulfured), alone or in salads. Make into apricot whip, add flaked nuts.	Good for anemia, constipation, and catarrh.
Artichokes Mineral Carbohydrate	Iodine Potassium Iron Silicon	Wash and steam. Use as cooked vegetable.	Good for soft bulk and minerals and general body builder.
Asparagus Mineral Carbohydrate	Calcium Iron Silicon	Cut tender portion from woody base. Remove scales if sandy. Cut up fine and steam.	Good for kidney and bladder disorders.
Avocado Mineral Fat	Chlorine Phosphorus Sulfur	Wash and peel. Eat alone, have in salads and soups. Good in sandwich filling. Goes well in any combination.	Body builder. Because of its non-irritating consistency it is good for colitis, ulcers. Patients can use as natural oil and bulk in intestines. Slightly laxative and good mineralizer for the body.
Banana Carbohydrate	Potassium Calcium Chlorine	Buy when spotted and no green tops. Wash. Eat alone or in salads, serve as a starch. Eat dead-ripe or baked.	Good for gaining weight. Used as natural bulk for irritated bowels, such as colitis, ulcers, or diarrhea.
Barley Carbohydrate Protein	Potassium Silicon	Use unpearled. Wash, steam, and serve as a starch, alone or in soups.	For gaining weight. Excellent for children up to ten years for silicon content.
Bass Protein	Phosphorus Chlorine Iodine	Broil, bake, or steam. Serve with natural sauces, or lemon.	Brain and nerve food. Use head, fins, and tail in broth for nerves and glands. Refer to Broths.

Food & Type	Predominant Chemical Elements	Best Way Prepared and Served for Digestion	Remedial Measures
Beans, Lima Carbohydrate Protein	Potassium Phosphorus Calcium Iron	Shell and wash fresh limas, steam, or use in vegetable and protein loaves.	Puréed for stomach ulcers. Good muscle-building food.
Beans, String Mineral Carbohydrate	Manganese Nitrogen	Wash, remove ends and strings. Cut once lengthwise and cut crosswise in one-inch strips. Steam.	Good body mineralizer.
Beef Protein	Phosphorus Potassium Chlorine	Should be broiled or roasted. Serve with geen vegetables and tomatoes or grapefruit.	Brain and nerve food. Good in anemia, especially for those over 20 years old, and for those who use up surplus energies.
Beets Mineral Carbohydrate	Potassium Fluorine Chlorine	Cut off leaves, leaving one-inch stems. Steam. Also shred and steam for variation.	Beet juice when combined with blackberry juice is a good blood builder. Use leaves like spinach.
Beet Greens Mineral Carbohydrate	Potassium Magnesium Iodine Iron	Clean and wash thoroughly. Use stems if tender. Cut up fine and steam like spinach.	Body mineralizer.
Blackberries Mineral Carbohydrate	Potassium Magnesium Iodine Iron	Wash and serve alone, with other fruit, or with protein.	Blood builder. Used for dysentery or diarrhea. Good for anemia.
Blueberries Mineral Carbohydrate	Potassium Calcium Magnesium	Wash and serve alone, with other fruit, or with protein	Blood purifier and body mineralizer.
Bread, Whole Wheat Protein Carbohydrate	Phosphorus Chlorine Calcium Silicon	To be eaten once a day with raw vegetable juices and salads. Sandwiches allowed but vegetable filling should be used.	When used discriminately, good for teeth, muscles, bones, and anemia.
Broccoli Mineral Carbohydrate	Potassium	Remove tough leaves, tough part of stalk. Wash thoroughly and steam.	Body mineralizer.

Food Remedy Troubleshooting Chart—*Continued*

Food & Type	Predominant Chemical Elements	Best Way Prepared and Served for Digestion	Remedial Measures
Brussels Sprouts Carbohydrate	Potassium Calcium Sulfur	Remove wilted leaves. Leave whole. Wash and soak in salt water 30 minutes. Steam.	Good mineralizer.
Butter, Cow Fat Mineral	Sodium Calcium Chlorine	Eaten on toast and served with cooked vegetables in modera- tion. Use sweet butter.	Good for eyes and supplying vitamin A, if not used in excess. Easiest fat to digest.
Buttermilk Mineral Protein	Sodium Calcium Chlorine	Best with citrus fruit or protein.	Good for diarrhea, gas, intestinal gas normal- izer, and for acidity.
Cabbage Mineral Carbohydrate	Potassium Sodium	Remove wilted outside leaves. Cut in fourths. Wash, soak in salt water. Boil seven minutes in uncovered pot. Also use raw in salad.	Good mineralizer.
Carrots Mineral Carbohydrate	Potassium Calcium Sulfur Silicon	Clean with vegetable brush. Shred fine, use in salads, raw, or steamed. A raw whole carrot daily develops children's teeth and jaws.	Eye food. Good for hair, nails. Easy to digest. One of the best foods to break a fast. Shred finely.
Casaba Mineral Carbohydrate	Potassium Sodium Chlorine Iron Silicon	Eat like other melons. Fill center with berries or sour cream. Good on hot afternoons.	Blood cleanser and cooler.
Cauliflower Mineral Carbohydrate	Potassium Calcium Sulfur Silicon	Remove leaves and woody base. Break flowers apart. Soak in salt water thirty min- utes. Steam.	Good intestinal cleanser.
Celery Mineral Carbohydrate	Chlorine Sodium Potassium Magnesium	Best eaten raw or in vegetable juice. May also be used steamed or in vegetable broth.	For arthritis, neuritis, rheumatism, acidity, high blood pressure, and for nerves. Use in juice form in good health and for every disease. Good blood cleanser.

Food & Type	Predominant Chemical Elements	Best Way Prepared and Served for Digestion	Remedial Measures
Chayote Mineral Carbohydrate	Potassium Magnesium Silicon	Wash, peel, cube, or slice, and steam.	Non-fattening and a good mineralizer.
Cheese, Cow *Cottage* Protein	Calcium Phosphorus Chlorine	Eaten as protein. Always serve with fruit and vegetables.	Hard to digest, but good source of complete protein. Dry or Farmer Style best.
Cheese, Goat *Cottage* Protein	Calcium Phosphorus Fluorine Chlorine	Always serve with fruit or vegetables.	Has fluorine in abundance. Good for bones, teeth, beauty, especially for children.
Cheese, Roquefort Protein	Calcium Phosphorus Fluorine Chlorine	Always serve with fruit or vegetables.	Has fluorine in abundance. Good for bones and teeth.
Cheese, Swiss Protein	Calcium Phosphorus Chlorine Sodium	Always serve with fruit or vegetables.	Good body builder.
Cherries, Wild *Black* Mineral Carbohydrate	Potassium Iron Magnesium	Eat alone or serve with protein.	For anemia, catarrh. Use one glass for three days in succession twice a month for chronic gall bladder trouble
Chervil Mineral Carbohydrate	Potassium Iron Phosphorus Sulfur	An herb eaten with salads, vegetables, protein, or carbohydrates	Body mineralizer.
Chicken Protein	Phosphorus Potassium Chlorine	Serve with non-starch vegetables and tomatoes or grapefruit	
Chicory Mineral Carbohydrate	Iron Sulfur Chlorine Potassium	A green to be served in salad.	Body mineralizer.
Chinese Cabbage Mineral Carbohydrate	Sodium Calcium Magnesium Iron	Serve raw or in salad, or prepared like cabbage.	Body mineralizer.

Food Remedy Troubleshooting Chart—*Continued*

Food & Type	Predominant Chemical Elements	Best Way Prepared and Served for Digestion	Remedial Measures
Chives Mineral Carbohydrate	Potassium Calcium Sulfur	Served in salads, with vegetables, or in cottage cheese.	Body mineralizer, good for catarrh, elimination.
Coconuts Protein Fat Mineral	Potassium Magnesium Phosphorus Chlorine	Milk and coconut meat eaten with fresh or diced fruit or vegetables.	Body builder and for weight building. Good for bones and teeth.
Corn Carbohydrate Protein	Potassium Phosphorus Silicon	Remove husk and silks with a stiff brush. Steam. Eat with green vegetables. Yellow corn better than white corn.	A great brain, bone, and muscle building food.
Cranberries Minerals Carbohydrate	Calcium Sulfur Chlorine	Eat with proteins.	Use as pack in rectum for hemorrhoids.
Cream, Cow Fat	Calcium Phosphorus Fluorine	Eat with fruit or vegetables.	Weight builder. Apply to chapped or sunburned skin.
Cucumbers Mineral Carbohydrate	Potassium Calcium Phosphorus Silicon Iron	Eaten in salad. Serve with a starch or protein.	Good for skin troubles, and for blood cooling.
Currants, Black Mineral Carbohydrate	Phosphorus Magnesium Potassium	Used as a sweet dried fruit; juice of fresh currants makes a refreshing drink.	Blood builder.
Dandelion Greens Mineral Carbohydrate	Potassium Calcium Manganese Chlorine	Discard greens with bud or blossoms as they are bitter. Cut off roots. Clean and wash thoroughly. Mix with sweet vegetables. Eat raw in salad or steam.	Cleanse liver and gall bladder. Body mineralizer.
Dates, Dry Carbohydrate	Chlorine	Wash, eat alone, or with sub-acid fruits or vegetables. Candy substitute.	Good for undernourishment.
Duck Protein	Potassium Phosphorus Chlorine	Broil or roast. Serve with green vegetables and grapefruit or tomatoes.	An easy protein to digest.

Food & Type	Predominant Chemical Elements	Best Way Prepared and Served for Digestion	Remedial Measures
Eggplant Mineral Carbohydrate	Potassium Phosphorus Chlorine	With protein or starch as a vegetable. Wash, steam, or bake whole, sliced or cubed. May be stuffed or used in roasts and loaves.	Good form of bulk. Good mineralizer.
Egg Yolk, Raw Mineral Fat Protein	Sulfur Chlorine Iodine Iron	Slowly cook, never fry, and serve with green vegetables, grapefruit, tomatoes, or fruit.	Excellent food for children. Brain, nerve, and gland food.
Endive Mineral Carbohydrate	Potassium Calcium Sulfur	Wash and serve in salads.	Body mineralizer.
Figs, Black Carbohydrate	Potassium Magnesium	Wash and eat alone, or with fruits. Good candy substitute.	A natural laxative. Good for constipation. Fig juice is an alternative drink when acid fruit juice cannot be taken.
Grapefruit, Fresh Mineral Carbohydrate	Sodium Potassium Calcium	Eaten alone, or with fruit or protein. Buy grapefruit when it has a brownish-yellow cast.	For fevers, reducing, blood cooling, and catarrh elimination.
Grapes Mineral Carbohydrate	Potassium Magnesium	Wash and serve alone or with other fruit or protein. Concord grapes are best.	Blood purifier. Grape diet once or twice every year should be taken. Good for intestinal cleansing. Especially good in all catarrhal conditions.
Halibut, Smoked Protein	Phosphorus Potassium Chlorine	Serve with green vegetables and grapefruit or tomatoes. Steam, bake, or broil.	Good source of complete protein. Good source of brain and nerve fat.
Honey Carbohydrate	Potassium Calcium Phosphorus	Because it is a concentrated sweet, use starches and green vegetables.	Honey in conjunction with onions makes good cough syrup when allowed to stand overnight. Eucalyptus honey is good for throat aliments.

Food Remedy Troubleshooting Chart—*Continued*

Food & Type	Predominant Chemical Elements	Best Way Prepared and Served for Digestion	Remedial Measures
Horseradish Mineral Carbohydrate	Sulfur Fluorine Potassium	Used in seasoning salads, salad dressings, sandwich filling and sauces.	Gall bladder and liver cleanser. Body mineralizer.
Kale Mineral Carbohydrate	Calcium Potassium	With green vegetables in salad. Wash, cut fine, and use raw, or in soups.	Green kale broth for supplying body calcium. Best source of calcium. Makes teeth and bones hard. Body mineralizer.
Kohlrabi Mineral Carbohydrate	Calcium Magnesium Potassium	Wash, peel, then cube, slice, or shred, and steam.	Body mineralizer.
Lamb Protein	Potassium Phosphorus Chlorine	Bake or broil. Serve with green vegetables and tomatoes or grapefruit.	Good source of protein. Brain, gland, nerve food.
Leeks Mineral Carbohydrate	Sodium Calcium	With green vegetables. Wash and use in salads.	Good for catarrhal conditions. Body mineralizer.
Lemons Mineral Carbohydrate	Calcium Magnesium Potassium	To be used alone as a drink or in salads served with a protein meal. Use instead of vinegar. Cuts sweetness of grape juice when added.	Catarrh elimination. Best used in fevers and liver disorders. Used in lime salts. Blood cooler and weight reducer. Good germicidal agent. Use as a skin bleach.
Lentils Protein Carbohydrate	Phosphorus Potassium	To be served with a green salad. Soak and cook until soft.	Muscle builder. Good when pureed for stomach ulcers and colitis.
Lettuce, Head Mineral Carbohydrate	Sodium Calcium Chlorine Potassium Iron	Wash well and use in salads. Green outside leaves are always best.	Slows digestion. Good for sleeplessness. In severe gas conditions, stop using in diet.
Lettuce, Romaine Mineral Carbohydrate	Calcium Sodium Potassium Chlorine	With green vegetables, in raw salads, with starches or proteins.	Mineralizer of the body.

Food & Type	Predominant Chemical Elements	Best Way Prepared and Served for Digestion	Remedial Measures
Lettuce, Sea Mineral Carbohydrate	Iodine Potassium Phosphorus Iron	Use powdered over salads, in drinks, or sprinkled on steamed vegetables.	Good source of iodine.
Limes Mineral Carbohydrate	Calcium Magnesium Potassium	To be used in a drink or on salads served with a protein meal.	Limes in whey, good as a blood cooler. Marvelous in congestion of the brain.
Mangoes Mineral Carbohydrate	Potassium Calcium Chlorine	Eaten like melons or served in salads.	Good for irritated intestinal disorders.
Milk, Cow Protein	Calcium Sodium Phosphorus	To be served with fruits. Served as a protein.	Complete protein. Use on eyes as a pack for inflammation.
Milk, Goat Protein	Sodium Fluorine Calcium Phosphorus	Use in place of cow's milk. Always have raw.	Better source of fluorine than cow's milk. Easier digested than cow's milk. Use raw.
Mushrooms Mineral Protein Carbohydrate	Potassium Phosphorus Iodine	Used as flavoring in meat substitutes, roasts, and in sauces.	Body mineralizer.
Muskmelon Mineral Carbohydrate	Sodium Potassium Silicon	Eat alone or with protein, or cut up in salads with other fruit.	Good mineralizer, blood cooler. Use instead of artificial soft drink.
Mustard Greens Mineral Carbohydrate	Sulfur Potassium Calcium Magnesium	Wash thoroughly, cut fine, and use in salads, or steam as a green vegetable. May be mixed with other greens.	Good body mineralizer, or source of calcium. Good liver and gall bladder cleanser.
Oats, Steel Cut Mineral Carbohydrate	Silicon Iodine Magnesium	Use with green vegetables, or raw salad. Must be well cooked. Soak before cooking.	Excellent children's food, especially when they lack silicon. Good source of silicon.
Okra Mineral Carbohydrate	Sodium Chlorine	Wash pods. Cut off stems. Use in broth and soups or steam. Serve separately with butter.	Good for stomach ulcers, irritated intestinal tract. Use in all broths for stomach disorders.

Food Remedy Troubleshooting Chart—*Continued*

Food & Type	Predominant Chemical Elements	Best Way Prepared and Served for Digestion	Remedial Measures
Olives Mineral Fat	Potassium Phosphorus	Serve with green vegetables, raw salad, or fruit.	Best source of potassium. Good brain and nerve food found in oil.
Onions, White Mineral Carbohydrate	Sulfur Potassium	Peel onions under water to keep eyes from watering. Serve cooked or raw in salads.	Good for all catarrhal, bronchial, and lung disorders.
Oranges Mineral Carbohydrate	Potassium Calcium Sodium Magnesium	To be used alone, with nuts, raw egg yolk, or with a protein meal.	Good to stir up acids, catarrhal settlements, and hard mucous.
Papaya Mineral Carbohydrate	Sodium Magnesium Sulfur Chlorine	Eat as a melon or serve in salads.	Good for stomach and intestinal disorders, especially the seeds made into a tea.
Parsnips Mineral Carbohydrate	Calcium Potassium Silicon	Wash, clean with stiff brush, cube, slice, or grate and steam.	Body mineralizer.
Parsley Mineral Carbohydrate	Calcium Potassium Sulfur Iron	Eaten raw with salads, meats, soups, and vegetables, Used as tea, and in raw vegetable juice.	Good for diabetes, for cleansing the kidneys, for controlling calcium in the body. Body mineralizer.
Peaches Mineral Carbohydrate	Calcium Phosphorus Potassium	Eaten alone, or in fruit salads with protein meal.	Good bowel regulator. Body mineralizer and blood builder.
Peanuts Protein Fat Carbohydrate	Phosphorus Silicon Potassium	Eaten with green leafy salad. Raw peanuts are best.	Hard to digest.
Pears Mineral Carbohydrate	Sodium Phosphorus	Eaten alone or in fruit salads with protein meals.	Good body mineralizer. Good intestinal regulator.
Peas, Garbanzo Protein Carbohydrate	Magnesium Phosphorus	Eaten as protein. Cook as dried beans, such as lentils and navy beans. Soak before cooking.	Good source of vegetable protein.

Food & Type	Predominant Chemical Elements	Best Way Prepared and Served for Digestion	Remedial Measures
Peas, Fresh Carbohydrate Mineral	Magnesium Calcium Chlorine	Shell and wash. Steam or use in broth. The pods also are good to use in broth with peas.	Body mineralizer.
Pecans Protein Fat	Phosphorus Calcium Potassium	Best eaten with green vegetables or fruit, or flaked on breakfast fruits.	Good nut protein. Used in weight building with celery and apples.
Persimmons Mineral Carbohydrate	Phosphorus Calcium	Eaten with other fresh fruit, protein, or alone.	Good body mineralizer. Good for irritable intestinal tract.
Pineapple Mineral Carbohydrate	Sodium Calcium Magnesium Iodine	Eaten alone, with other fresh fruit, as in salad, or with protein.	Good for sore throat, catarrhal conditions, good blood builder, aids digestion.
Plums Mineral Carbohydrate	Magnesium	Eaten with other fresh fruit, alone, or with protein.	Good laxative and bowel regulator.
Pomegranate Mineral Carbohydrate	Sodium Magnesium	Squeeze out juice and drink very fresh.	Pomegranate juice with whey is good in brain and nerve congestion, and is a blood cleanser. Pomegranate juice is beneficial in bladder complaints.
Popcorn Carbohydrate	Phosphorus	May be eaten with green leafy vegetable salad and a cream dressing.	Good for intestinal roughage.
Potato, Baked Mineral Carbohydrate	Potassium Phosphorus Magnesium Silicon	Clean with stiff brush. Parboil two minutes. Butter skins and bake in slow oven.	Best source of starch. Use potato peeling in broths. Use for poultices.
Prunes Mineral Carbohydrate	Potassium Phosphorus Magnesium	Wash, place in clean water, bring to boil, and let stand overnight. Use as dried fruit for breakfast, whipped for a dessert, or in salads.	Good bowel regulator. Good source of nerve salts.
Pumpkin Carbohydrate	Sodium Iron Phosphorus	Eat with vegetable meal. Can be made into custards.	Body builder.

Food Remedy Troubleshooting Chart—*Continued*

Food & Type	Predominant Chemical Elements	Best Way Prepared and Served for Digestion	Remedial Measures
Radishes, Black Mineral Carbohydrate	Potassium Phosphorus Magnesium	Use as seasoning.	Has Raphanon which is extremely good in gall bladder and liver disorders.
Radishes, Red Mineral Carbohydrate	Potassium Phosphorus Magnesium	Use raw in salads, with green vegetables and starches.	Good source of sulfur. Good for catarrh.
Raisins Mineral Carbohydrate	Potassium Phosphorus Chlorine	With vegetables, starch, or protein. Wash well. Soak. Use in cereals for sweetening or in salads.	Concentrated sweet. Good body builder and good energy food.
Raspberries Mineral Carbohydrate	Sodium Iron	Wash well. Serve alone or with fruit or protein.	Blood mineralizer. Neutralizes acidity. Good for anemia.
Rice, Natural Brown Carbohydrate	Phosphorus Sodium	Steam and serve with green vegetables.	Good body building food. Good for bones, teeth, etc.
Rye, Whole Carbohydrate	Phosphorus Magnesium Silicon	Use with raw green vegetables.	Good source of silicon.
Spinach Mineral Carbohydrate	Potassium Silicon	Cut off roots and dead leaves. Wash, cut fine, and use raw or steamed.	Body mineralizer.
Squash Carbohydrate Mineral	Sodium Magnesium	Cut into pieces, or leave whole, and bake or steam.	Body builder, and bowel regulator.
Strawberries Mineral Carbohydrate	Calcium Sodium	Wash, and use fresh with or without fruit or protein.	Acid neutralizer when eaten ripe.
Swiss Chard Mineral Carbohydrate	Sodium Calcium Magnesium Iron	Wash thoroughly. Cut up in one inch pieces. Steam. Tender sections may be used raw in salads.	Body mineralizer.
Tomatoes Mineral Carbohydrate	Potassium Sodium Chlorine	Use only ripest tomatoes. Use in salads, broths, or steamed. Use with proteins.	Consider canned tomatoes best. Always use with a protein. Use also in packs and poultices.

Food & Type	Predominant Chemical Elements	Best Way Prepared and Served for Digestion	Remedial Measures
Turnips Mineral Carbohydrate	Potassium Calcium	Wash and shred. Use raw in salads or serve steamed.	Body builder. White turnip juice good for asthma, sore throat, and bronchial disorders.
Turnip Leaves Mineral Carbohydrate	Calcium Magnesium	Serve raw, in salads in vegetable juices, or steam with greens.	Good for controlling calcium in the body.
Walnuts Protein Fat	Manganese Phosphorus Magnesium	To be used with fruit or vegetables in salads.	Black walnuts are best source of brain and nerve manganese food.
Watercress Mineral Carbohydrate	Sulfur Chlorine Calcium	Wash well and use as salad green or garnish.	Body mineralizer.
Watermelon Mineral Carbohydrate	Silicon Calcium Sodium	Use in fruit salads and protein meals, or serve alone.	Good for kidneys. Blood cooler, and good source of silicon.
Wheat, Whole Carbohydrate	Phosphorus Silicon	Used in breads and cereals. Chew well, because starches must be mixed well with saliva in mouth to be digested properly.	Body builder of bone and teeth, especially for children. Best used in dry forms to promote vigorous chewing.
Whey Mineral Protein	Sodium Calcium Chlorine	Add fruit juices to whey and drink two or three times daily between meals, drink alone, or with meals.	Good source of mineral salts. Easy to digest, good blood builder. Important culture for the friendly bacteria in the intestinal tract.
Zucchini Mineral Carbohydrate	Potassium	Wash, cut into pieces. Steam as vegetable or cut up raw in salads.	Body mineralizer.

Note: For a detailed breakdown of information contained in this abbreviated table, see Part Two of this book.

Conclusion

Making Changes

P lato, the Greek philosopher who lived at the same time as Hippocrates, believed that the test of truly knowing what is right is to act on what you know. You can certainly put that bit of philosophy to work right now with *Foods That Heal*. When you do, I am confident that you will have proven to yourself that right nutrition works.

Understanding and applying the principles in this book is your first step toward a more satisfying life. In these pages, you have seen how the relationship between food and health has been known for more than 2,000 years. But both the population at large and most health professionals have resisted putting sound nutritional knowledge into practice until only very recently. By acting on what you now know, you can be a tradition breaker who puts good health back in the forefront of our culture where it belongs.

Luckily, America is growing more interested in natural health. This interest is fruit of the seed planted around the year 431 B.C. by none other than Hippocrates himself. The seed was watered, fertilized, and nurtured over the years by a few brave souls who did not mind being ignored, neglected, ridiculed, or mocked by the great majority of people in their own time. Why? Because they believed they were carrying a banner of truth that would someday be acknowledged. They were the health nuts of history. To me they are heroes.

In my imagination, I would like to think that those special

people who believed in foods that heal survived the plagues and epidemics that devastated large portions of the world. And because they ate healthfully, their immune systems were strong enough to throw off even the most potent onslaughts history hurled their way: bubonic plague, cholera, typhus, and smallpox.

But simply surviving plagues and epidemics is not enough. The good news about food is that it can enrich our lives and keep us happy. We should therefore be highly motivated to seek health and well-being as the Number One priority in the world today.

We should prepare for a new day in health. And perhaps that day is not far away. I certainly detect a positive change in consciousness among the people who have been attending my classes and lectures over the past few years.

Today, the value of nutrition is being affirmed by more well-educated and respected health professionals than ever before. In addition, people from all walks of life are waking up to the common sense that we are what we eat. As a result, the climate of acceptance is more favorable.

Experience has shown me that people will listen to what you have to say if you are a good advertisement for what you're selling. If your level of wellness attracts their attention and strikes their interest, you're on your way to proving that good nutrition works. People will wonder if they can become as healthy and dynamic as you and, when they get up the courage, they'll ask you how you did it.

Give them the good news.

Part Two

A Guide to Fruits and Vegetables

Listing

Fruits and Vegetables

I have spent many years collecting the information that appears in this guide. I hope the information will be valuable to you. This guide to foods that heal shows how to use foods as medicine, in the tradition of V. G. Rocine and Hippocrates.

Each fruit or vegetable is described in considerable detail, and the therapeutic value of the food is given. Not every fruit or vegetable is in this guide. I have only included the ones for which I could get all the information I wanted to share with you. Perhaps one of you will finish what I have started!

In each food description, the nutritional value of each food is broken down. Not all of the vitamins and minerals are mentioned in the nutrient contents lists, but if the food contains a sufficient amount of one of these nutrients—potassium, for example—I have mentioned it in the text.

In the list of nutrients (titled Nutrients in One Pound), a dash sometimes appears after a nutrient. This means that there is a lack of reliable data for this particular nutrient in this food. If a zero appears after the nutrient, then the amount of this nutrient is either too small to measure or does not appear at all in this food. If the word *trace* appears after a nutrient, then that nutrient is present in quantities less than a standard limit.

The amount of each vitamin or mineral is expressed in grams (g) or milligrams (mg). The amount of vitamin A is always expressed in International Units (I.U.).

MY FOOD LAWS

I have worked successfully with thousands of patients. My main approach to patient care is to use a balanced diet and to make sure the basic food laws are followed. (See My Food Laws in Chapter 3 in Part One.) To allow for individual deficiencies and needs, I advise certain supplements or specialty foods in addition to the main diet.

When a balanced diet and my food laws are followed, nutritional value is guaranteed, especially when quality foods are used. At my ranch, we raise most of our fruits and vegetables organically, so the quality of the food is high. The average person will get adequate amounts of most vitamins and minerals, because the overall food regimen will provide the balance and variety needed. Enzymes, prostaglandins, chlorophyll, and fiber are provided when these food laws are followed, and are necessary for the best of health.

RESEARCH RESULTS

A four-year study published in 1988 showed that health care costs for obese persons averaged 11 percent higher than for those who are at their proper weight. People who walk less than half a mile a week have 14 percent more health care claims than those who walk at least a mile and a half each week. Another study showed that two out of ten Americans who said they felt well actually had at least one chronic disease. The National Academy of Sciences has linked diet, exercise, and other lifestyle components (such as smoking) with cancer and heart disease. These diseases claim thousands of lives a year. What should we learn from studies like these?

One thing we can conclude is that we must learn more about foods and health. Also, we should start learning how to motivate ourselves to maintain a healthy discipline, especially with regard to food and exercise.

Once, when I had finished giving a lecture, an elderly lady came to me and said, "I had a salad last week and I don't feel a bit better." What we all must know is that it isn't what we do once in a while that has an impact on our health—we must follow a diet of nutrition and exercise every day.

Another thing we can conclude is that our diet has a direct effect on our health. In the past forty years in this country, there has been a major trend toward eating more meat and dairy products and less fruits and vegetables. We also eat more meals away from home than ever before—especially in fast food restaurants. Instead of eating a diet that primarily consists of meat and dairy products, we should begin to eat more fruits and vegetables. Then, our health will improve.

Although I eat some meat now, I was a vegetarian for seventeen years. I believe in the vegetarian way of life. The vegetarian diet is pure and clean, but it isn't for everyone. It doesn't usually work for those who have a demanding job with a lot of stress, or for those in extremely competitive occupations. If you use up a lot of adrenaline in your way of life, then the vegetarian diet will not accommodate your needs. Yet, groups and cultures that use a great deal of fruits and vegetables in their diets are interestingly different from other cultures, especially when it comes to measuring health levels.

For example, members of the Seventh-Day Adventist Church place an unusually strong emphasis on following a healthy regimen. A relatively high proportion of them faithfully follow the guidelines recommended by their church. They eat more fruits, vegetables, whole grains, nuts, and seeds than most people eat. Researchers have found that the Seventh-Day Adventists have a far lower incidence of heart disease, cancer, and many other chronic diseases than those who follow a typical American diet.

EAST AFRICAN VILLAGERS

Dr. Denis P. Burkitt, a British surgeon, studied rural East African villagers for many years. He found that problems like obesity, diabetes, hiatus hernia, appendicitis, diverticulosis, colitis, polyps, and cancer of the colon simply didn't exist among these people. He began to look for the reason.

Dr. Burkitt discovered that the diet of these rural East Africans was high in fresh fruit, vegetables, and coarse-ground cereal grains. The bran from the grain is highly water absorbent, increasing the bulk in the bowel and speeding up elimination time. Fruit and vegetables, which are high in fiber, also help in this process.

There are several benefits to speeding up bowel transit time. Gas and putrefaction are reduced. More cholesterol and fats are eliminated from the body. The amount of toxins absorbed through the bowel wall is greatly lowered. The bowel wall is kept cleaner, and its musculature is stronger and more responsive.

Dr. Burkitt found that the bowel transit time for a rural Bantu native was more than twice as fast as that of the average Englishman. The length of time food wastes remain in the bowel determines three things:

1. How putrefactive they become.
2. How much the undesirable bacteria multiply.
3. The amount of time toxins are able to be in contact with the bowel wall.

The increased health risks are obvious. Also, the Bantu native would absorb less than half of the cholesterol that is assimilated through the bowel wall of the average Englishman.

Dr. Burkitt also credits the absence of products like white sugar, white flour, and other refined carbohydrates for the health benefits enjoyed by the East African villagers. In contrast, those who moved to the cities to work usually adopted the urban diet, and quickly began manifesting the disease problems that are so common in just about all "civilized" environments.

A few other observations are in order here. Most likely, the fruits, vegetables, and grains grown by East African villagers are free of pesticides. The soil they are grown in has not been damaged by the addition of petrochemical fertilizers. There is no serious problem, in that part of the world, with acid rain. Most villagers get a great deal of exercise by walking or running. These factors would all have a positive influence on their health statistics.

FRUIT AND VEGETABLE FACTS

Before you read this guide, there are a few basic things that you should know about fruits and vegetables. Generally, vegetables are considered a better source of minerals, while fruit is considered a better source of vitamins. Taking into account the fact that most fruit in this country is picked before it is ripe for shipping to market, vegetables would probably be superior on both counts, since unripe fruit is lower in vitamin content.

Unlike vegetables, fruit is ripened by the sun, which is a sodium star. Oddly enough, the sweetness of fruit is influenced by its sodium content as well as its fruit sugar content. Lemons, for example, have more sugar per equal weight than oranges, although oranges taste sweeter. (I really don't like to use citrus fruit in the diets I design, because it is picked six weeks before it would normally be ripe. Green citrus only adds to and stirs up the acidity in the body.)

Ripe fruit is best to eat, if you can get it. Dried fruit is all right, but you should realize that in the drying process, the sugar has become highly concentrated. It is best to reconstitute dried fruit before using it by soaking it in boiling water. This will also kill insect eggs.

Green vegetables are a wonderful source of vitamin A, in the form of carotene. The highest vegetable sources of vitamin A are carrot juice and hot chili peppers, which both contain over 10,000 International Units per hundred grams. If we consider chlorella a vegetable, it has a vitamin A content of nearly 50,000 I.U. per hundred grams. Chlorella, which is a green, single-celled edible alga, is also high in chlorophyll, vitamin B_{12}, and nucleic acids, which are sometimes called ''youth factors.''

I believe that fruits and vegetables should be eaten seasonally. Most fruits are harvested in summer and fall, and during these seasons we should be building up a reservoir of vitamins for the winter. More toxins are eliminated during these seasons: fruit cleanses the body by stimulating bowel activity, and in the summer we wear lighter clothing and perspire more, increasing our elimination of toxins.

In the winter, we can eat reconstituted dried fruit. I use two large fruit driers to preserve summer fruit. I do not recommend sulfured dried fruit, because the chemical sulfur can cause skin and bowel problems. Home-canned fruit can also be used in the winter, but its nutritional value is lower than that of dried or frozen fruit, in most cases.

Vegetables have a lower sugar and moisture content, so they keep for a much longer period of time. People who live in cold climates are often able to keep potatoes, turnips, yams, winter squash, and other vegetables very well. Usually they store these vegetables in dry storage bins.

As cold approaches, many people have an appetite for heat-producing foods. Starches such as potatoes, whole cereal grains, yams, and some species of squash are good heat producers. Kale and barley soup is also a heat-producer, as well as a good source of calcium and vitamin A.

Nutrition is a science that is still in its infancy. There is still so much to discover. People are finally realizing that a knowledge of nutrition is necessary. This knowledge will not only enhance our well-being, but will enable us to survive in an era that is filled with health hazards.

APPLE

One of the first things a child learns is the alphabet, and almost always, "A is for apple." The apple has been around for so long that it can be called the first fruit. Hieroglyphic writings found in the pyramids and tombs of the ancient Egyptians indicate that they used the apple as both a food and a medicine. It not only has been at the beginning of alphabet songs, but has been the center of legends, folklore, and even religion, for thousands of years, from Adam and Eve to Johnny Appleseed.

The people of the United States love apples. The state of Washington produces 32,000,000 boxes of apples a year. Washington's orchards supposedly began from a single tree that was planted in 1827 from a seed given to Captain Simpson of the Hudson Bay Company by a young woman from London. That tree is still standing!

Years ago, apples were used to relieve gout, bilious constitutions, skin eruptions, and nerves. They are so popular around the world that they have all kinds of superstitions and traditions attached to them. The peasants of Westphalia used apples mixed with saffron as a cure for jaundice. There is also a legend in Devonshire, England, that an apple rubbed on a wart will cure it. On Easter morning, peasants in a province of Prussia ate an apple to insure against fever. The Turks gave the apple the power of restoring youth.

There are so many varieties of apples that almost anyone can find an apple to suit his palate. Since there are summer, winter, and fall varieties, apples can be had fresh all year around.

Today, doctors use apple therapy for stubborn cases of diarrhea in patients of all ages, including babies. Raw apple is scraped in very fine slices or used in a specially prepared concentrate. This treatment is often used for what is called the "lazy colon," and is also good for babies who are ready to begin a solid diet. Because so many of the essential vitamins and minerals in apples contain a predigested form of fruit sugar, it is an ideal fruit for infants and invalids.

When you cook apples, be sure to do so over a very low flame. It is best to cook them in a stainless steel utensil, so that the delicate pectin, vitamins, and minerals will be preserved as much as possible. Apples, of course, are best raw and are good in various kinds of salads.

THERAPEUTIC VALUE

Apples are an alkaline food. They are also an eliminative food, and contain pectin, which has the ability to take up excess water in the intestines and make a soft bulk that acts as a mild, nonirritating stimulant. This stimulant helps the peristaltic movement and aids in natural bowel elimination.

The iron content of the apple is not high, but it has a property that helps the body absorb the iron in other foods, such as eggs and liver. It does contain a generous amount of calcium, and this calcium aids the system in absorbing the calcium in other foods.

Apples contain 50 percent more vitamin A than oranges. This vitamin helps ward off colds and other infections and promotes growth. It also keeps the eyes in good condition, and prevents night blindness.

Apples have an abundant supply of vitamins. They contain more vitamin G than almost any other fruit. This is called the "appetite vitamin," and promotes digestion and growth. They are rich in vitamin C, which is a body normalizer and is essential in keeping bones and teeth sound. The vitamin that is so important in maintaining nerve health, vitamin B, is also found in apples.

Apples are good for low blood pressure and hardening of the arteries because they are powerful blood purifiers. They also benefit the lymphatic system.

The juice of apples is good for everyone. It can be used in a cleansing and reducing diet, but speeds up bowel action, and can produce gas if bowels are not moving well. Apple juice or concentrate added to water makes a solution that heals bowel irritation when given as an enema.

Raw apples should be used for homemade apple juice, which should be consumed immediately after preparation. Save the peelings for health tea, which is excellent for the kidneys. This tea is simply made from steeped apple peelings. It is especially tasty when a little honey has been added to it.

NUTRIENTS IN ONE POUND

Calories	258	Iron	1.2 mg
Protein	1.2 g	Vitamin A	360 I.U.
Fat	1.6 g	Thiamine	.15 mg
Carbohydrates	59.6 g	Riboflavin	.08 mg
Calcium	24 mg	Ascorbic acid	18 mg

APRICOT

The apricot is said to have originated in China. It spread from there to other parts of Asia, then to Greece and Italy. As early as 1562 there is mention of the apricot in England in Turner's *Herbal*.

It is recorded that the apricot grew in abundance in Virginia in the year 1720. In 1792 Vancouver, the explorer, found a fine fruit orchard that included apricots at Santa Clara, California. The fruit was probably brought to California by the Mission Fathers in the eighteenth century.

The apricot is a summer fruit, and is grown in the Western United States. California produces 97 percent of the commercial apricot crop. Only about 21 percent of the apricots produced commercially are sold fresh; the remainder are canned, dried, or frozen.

Tree-ripened apricots have the best flavor, but tree-ripened fruit is rarely available in stores, even those close to the orchard. The next best thing to a well-matured apricot is one that is orange-yellow in color, and plump and juicy. Immature apricots never attain the right sweetness or flavor. There are far too many immature apricots on the market. They are greenish-yellow, the flesh is firm, and they taste sour. Avoid green and shriveled apricots.

THERAPEUTIC VALUE

Apricots may be eaten raw in a soft diet. Ripe apricots are especially good for very young children and for older people. This fruit is quite laxative, and rates high in alkalinity. Apricots also contain cobalt, which is necessary in the treatment of anemic conditions.

Apricots may be puréed for children who are just beginning to eat solid foods. Apricot whip for dessert is wonderful, and apricots and cream may be used in as many ways as possible. They make good afternoon and evening snacks.

Dried apricots have six times as much sugar content as the fresh fruit. Therefore, persons with diabetic conditions must be careful not to eat too much dried apricot. Because of its sugar content, however, it is good when we need an energy boost.

Dried fruits should be put in cold water and brought to a boil the night before, or permitted to soak all night, before eating. Bringing the water to a boil kills any germ life that may be on the fruit. Sweeten only with honey, maple syrup, or natural sugars.

NUTRIENTS IN ONE POUND

Calories	241	Iron	2.1 mg
Protein	4.3 g	Vitamin A	11,930 I.U.
Fat	0.4 g	Thiamine	0.13 mg
Carbohydrates	55.1 g	Riboflavin	0.17 mg
Calcium	68 mg	Niacin	3.2 mg
Phosphorus	98 mg	Ascorbic acid	42 mg

ARTICHOKE

The artichoke is believed to be native to the area around the western and central Mediterranean. The Romans were growing artichokes over 2,000 years ago, and used it as a green and a salad plant.

Artichokes were brought to England in 1548, and French settlers planted them in Louisiana in the mid-nineteenth century. California is now the center of the artichoke crop, and its peak season is March, April, and May.

The name "artichoke" is derived from the northern Italian words "articiocco" and "articoclos," which refer to what we know to be a pine cone. The artichoke bud does resemble a pine cone.

There is a variety of vegetable called the Jerusalem artichoke, but it is not a true artichoke. It is a tuberous member of the sunflower family. Here, we refer to the two types of true artichokes, the Cardoon (cone-shaped) and the Globe. The most popular variety is the Green Globe.

The artichoke is a large, vigorous plant. It has long, coarse, spiny leaves that can grow to three feet long. The artichoke plant may grow as high as six feet tall.

A perennial, the artichoke grows best in cool, but not freezing, weather. It likes plenty of water, and rain and fog, so is best suited to the California coast, especially the San Francisco area.

For a good quality artichoke, select one that is compact, plump, and heavy, yields slightly to pressure, and has large, tightly clinging, fleshy leaf scales that are a good color. An artichoke that is brown is old or has been injured. An artichoke is overmature when it is open or spreading, the center is fuzzy or dark pink or purple, and the tips and scales are hard. March, April, and May are the months when the artichoke is abundant.

The parts of the artichoke that are eaten are the fleshy part of the leaves and heart, and the tender base. Medium-sized artichokes are best—large ones tend to be tough and tasteless. They may be served either hot or cold, and make a delicious salad.

To prepare artichokes, cut off the stem and any tough or damaged leaves. Wash the artichoke in cold running water, then place in boiling water, and cook twenty to thirty minutes, or until tender. To make the artichoke easier to eat, remove the choke in the center, pull out the top center leaves, and, with a spoon, remove the thistle-like inside.

To eat artichokes, pull off the petal leaves as you would the petals of a daisy, and bite off the end.

THERAPEUTIC VALUE

Artichoke hearts and leaves have a high alkaline ash. They also have a great deal of roughage, which is not good for those who have inflammation of the bowel. They are good to eat on a reducing diet.

Artichokes contain vitamins A and C, which are good for fighting off infection. They are high in calcium and iron.

NUTRIENTS IN ONE POUND (including inedible parts)

Calories	60	Iron	2.4 mg
Protein	5.3 g	Vitamin A	290 I.U.
Fat	.4 g	Thiamine	.14 mg
Carbohydrates	19.2 g	Riboflavin	.09 mg
Calcium	93 mg	Niacin	1.7 mg
Phosphorus	160 mg	Ascorbic acid	22 mg

ASPARAGUS

The ancient Phoenicians brought asparagus to the Greeks and Romans. It was described in the sixteenth century by the English writer Evelyn as ''sperage,'' and he said that it was ''delicious eaten raw with oyl and vinegar.''

When selecting asparagus, choose spears that are fresh, firm, and tender (not woody or pithy), with tips that are tightly closed. Watch for signs of decay, such as rot and mold. If the tip of the spear appears wilted, the asparagus is really too old to be good. From the tip to all but an inch of the base, the stalk should be tender. Angular stalks indicate that they are tough and stringy.

Store asparagus wrapped in a damp cloth or waxed paper, and keep refrigerated until you are ready to use it. Asparagus loses its edible quality when it is subjected to dryness and heat, which reduce the sugar content and increase the fiber content.

Asparagus is a perennial herb, and is a member of the Lily of the Valley family. It can be served hot, with drawn butter; cold, in a salad; in soups; and as a sandwich filling or flavoring.

The season for asparagus is February through July, and the peak months are April, May, and June. Early spring asparagus is from California; late spring asparagus is shipped in early April or late May from Maryland, Delaware, New Jersey, Pennsylvania, Massachusetts, Michigan, Illinois, and Iowa.

Green asparagus is the most nutritious. Some varieties are green-tipped with white butts, and some are entirely white. Most of the white variety is canned.

Asparagus is best when cooked in stainless steel, on low heat. This leaves the shoots tender and retains their original color. If cooked with the tips up, more vitamin B_1 and C will be preserved. The liquid can be saved and used in vegetable cocktails.

THERAPEUTIC VALUE

Asparagus acts as a general stimulant to the kidneys, but can be irritating to the kidneys if taken in excess or if there is extreme kidney inflammation. Because it contains chlorophyll, it is a good blood builder.

Green asparagus tips are high in vitamin A, while the white tips have almost none. This food leaves an alkaline ash in the body. Because they have a lot of roughage, only the tips can be used in a soft diet. They are high in water content and are considered a good vegetable in an elimination diet.

Many of the elements that build the liver, kidneys, skin, ligaments, and bones are found in green asparagus. Green asparagus also helps in the formation of red blood corpuscles.

NUTRIENTS IN ONE POUND

Calories	90	Iron	3.11 mg
Protein	7.5 g	Vitamin A	3,430 I.U.
Fat	0.7 g	Thiamine	0.54 mg
Carbohydrates	13.1 g	Riboflavin	0.59 mg
Calcium	71 mg	Niacin	3.9 mg
Phosphorus	211 mg	Ascorbic acid	113 mg

AVOCADO

There are over 400 varieties of avocado. Some have smooth skin and are green, and some are rough and black. The avocado is considered a neutral fruit, because it blends well with almost any flavor and mixes well with either vegetables or fruit.

The avocado came from Persia. It has been popular in South America, Central America, and Mexico for centuries. The ancient Aztecs left evidence that the avocado was in their diet, as did the Mayans and Incas. It is known that the avocado was eaten by Jamaicans in the seventeenth century. This fruit grows wild in tropical America today, but is primarily grown as a crop in southern California.

THERAPEUTIC VALUE

Avocado at its peak contains a high amount of fruit oil. Fruit oil is a rare element, and it gives avocado its smooth, mellow taste and nut-like flavor. Fruit oil also gives the avocado its high food energy value. Unlike most fruit, it contains very few carbohydrates.

The avocado contains fourteen minerals, all of which regulate body functions and stimulate growth. Especially noteworthy are its iron and copper contents, which aid in red blood regeneration and the prevention of nutritional anemia. It also contains sodium and potassium, which give this fruit a high alkaline reaction.

The avocado contains no starch, little sugar, and has some fiber or cellulose.

NUTRIENTS IN ONE POUND

Calories	568	Iron	2.0 mg
Protein	7.1 g	Vitamin A	990 I.U.
Fat	55.8 g	Thiamine	.37 mg
Carbohydrates	21.4 g	Riboflavin	.67 mg
Calcium	34 mg	Niacin	5.4 mg
Phosphorus	143 mg	Ascorbic acid	48 mg

BANANA

Bananas were cultivated in India 4,000 years ago. In 1482, the Portuguese found the banana on the Guinea coast and carried it with them to the Canary Islands. Spanish priests are credited with having introduced this fruit to tropical America when they arrived as missionaries in the sixteenth century. Now, the banana can be found in all tropical countries.

The first known species of banana is the plaintain, or cooking banana. The plaintain has a salmon-colored pulp, a cheesy, gummy texture, and a slightly acid taste. This fruit has been a substitute for bread or potatoes in many countries, and is slowly being introduced to the United States.

Bananas are usually harvested green, shipped green, and ripened by wholesale fruit jobbers in air-conditioned ripening rooms. The Gros Michel variety is the most popular of the many varieties. It produces the largest and most compact bunch, which makes it easier to ship. The thick skin of the banana protects the soft fruit.

Other popular varieties of banana are the Claret, or red banana, which has a gummy flesh; the Lady Finger, which is the smallest variety, but has a delicate, sweet flavor; and the Apple, which has an acid flavor and tastes somewhat like a mellow apple.

In the tropics, bananas are often cooked and served with beans, rice, or tortillas. In the Latin American countries, the ripe banana is sometimes dried in the sun in much the same manner as figs and raisins. They are often sliced when ripe and left in the sun until they are covered with a coating of white, sugary powder that arises from their own juices.

The banana has no particular growing season. A ripe banana is firm, with a plump texture, strong peel, and no trace of green on the skin. A skin that is flecked with brown means the fruit is good.

Fully ripe bananas are composed of 76 percent water, 20 percent sugar, and 12 percent starch.

THERAPEUTIC VALUE

The sugars in the banana are readily assimilated, and they contain many vitamins and minerals, and a great deal of fiber. They are excellent for young children and infants and are good in reducing diets because they satisfy the appetite and are low in fat.

Because they are so soft, they are good for persons who have intestinal disturbances, and for convalescents. Bananas feed the natural acidophilus bacteria of the bowel, and their high potassium content benefits the muscular system.

NUTRIENTS IN ONE POUND

Calories	299	Iron	1.8 mg
Protein	3.6 g	Vitamin A	1,300 I.U.
Fat	0.6 g	Thiamine	0.27 mg
Carbohydrates	69.9 g	Riboflavin	0.19 mg
Calcium	24 mg	Niacin	1.7 mg
Phosphorus	85 mg	Ascorbic acid	29 mg

BEET

The beet has been cultivated for its roots and leaves since the third or fourth century B.C. It spread from the area of the Mediterranean to the Near East. In ancient times it was used only for medicinal purposes—the edible beet root we know today was unknown before the Christian era.

In the fourth century beet recipes were recorded in England, and in 1810 the beet began to be cultivated for sugar in France and Germany. It is not known when the beet was first introduced to the United States, but it is known that there was one variety grown here in 1806.

Sugar beets are usually yellowish-white, and are cultivated extensively in this country. The garden beet ranges from dark purplish-red to a bright vermillion to white, but the most popular commercial variety is red.

Beets are available in the markets all year. Their peak season is May through October. They are primarily grown in the southern United States, the Northeast, and the west coast states.

When selecting beets, do not just look at the condition of the leaves. Beets that remain in the ground too long become tough and woody, and can be identified by a short neck, deep scars, or several circles of leaf scars around the top of the beet.

THERAPEUTIC VALUE

Beets are wonderful for adding needed minerals. They can be used to eliminate pocket acid material in the bowel and for ailments in the gall bladder and liver. Their vitamin A content is quite high, so they are not only good for the eliminative system, but also benefit the digestive and lymphatic systems.

NUTRIENTS IN ONE POUND (without tops)

Calories	147	Iron	3.4 mg
Protein	5.4 g	Vitamin A	22,700 I.U.
Fat	0.3 g	Thiamine	0.07 mg
Carbohydrates	32.6 g	Riboflavin	0.16 mg
Calcium	51 mg	Niacin	1.5 mg
Phosphorus	92 mg	Ascorbic acid	80 mg

BLACKBERRY

Blackberries are native to both North America and Europe, but cultivation of this fruit is largely limited to North America. In the early days of the United States, when land was cleared for pasture, blackberry bushes began to multiply. There are many hybrids of blackberries, and both man and nature have had a hand in this process. By 1850, cultivated blackberries had become very popular.

Blackberries are now cultivated in almost every part of the United States. Texas and Oregon probably have the largest numbers of acres planted with blackberries. Cultivation of this berry has been slow, because wild berries grow in abundance all over the country. The summer months are the peak season for blackberries.

A quality berry is solid and plump, appears bright and fresh, and is a full black or blue color. Do not choose berries that are partly green or off-color, because the flavor will not be good.

THERAPEUTIC VALUE

Blackberries are high in iron, but can cause constipation. They have been used for years to control diarrhea. If blackberry juice is mixed with cherry or prune juice, the constipating effect will be taken away. If one can take blackberry juice without constipating results, it is one of the finest builders of the blood.

NUTRIENTS IN ONE POUND

Calories	294	Iron	4.1 mg
Protein	5.4 g	Vitamin A	1,460 I.U.
Fat	3.6 g	Thiamine	0.12 mg
Carbohydrates	59.9 g	Riboflavin	0.3 mg
Calcium	163 mg	Niacin	1.3 mg
Phosphorus	154 mg	Ascorbic acid	106 mg

BLUEBERRY

Blueberries originally grew wild in North America, and in many places they still do. By 1910 there were at least two varieties being cultivated for market. Breeding and selection have made these berries popular, but wild fruit is also marketed.

Blueberries are available from early May through August, and the peak month is July. Canada and the northeastern United States produce the greatest amount of blueberries, because they grow best when the days are long and the nights cool. In any one area the blueberry season usually lasts from six to seven weeks.

Quality blueberries are plump, look fresh, clean, and dry, are fairly uniform in size, and are a deep blue, black, or purplish color. Overripe berries are dull in appearance, soft and watery, and moldy.

THERAPEUTIC VALUE

Blueberries contain silicon, which helps rejuvenate the pancreas. They are said to be good for diabetic conditions.

NUTRIENTS IN ONE POUND

Calories	310	Iron	3.6 mg
Protein	2.9 g	Vitamin A	420 I.U.
Fat	2.1 g	Thiamine	————
Carbohydrates	63.8 g	Riboflavin	————
Calcium	63 mg	Niacin	————
Phosphorus	54 mg	Ascorbic acid	58 mg

BROCCOLI

Broccoli was grown in France and Italy in the sixteenth century, but was not well known in this country until 1923, when the D'Arrigo Brothers Company made a trial planting of Italian sprouting broccoli in California. A few crates of this were sent to Boston, and by 1925 the market was well established. Since then, the demand for broccoli has been steadily on the increase.

Broccoli is a member of the cabbage family. California, Arizona, and Texas are the main broccoli-producing states.

When choosing broccoli, look for tenderness in the stalk, especially the upper portion. If the lower portion of the stalk is tough and woody, and if the bud clusters are open and yellow, the broccoli is overmature and will be tough. Fresh broccoli does not keep, so purchase only as much as you can immediately use.

Broccoli is often gas-forming, but if cooked in a steamer or over a very low fire, this may be avoided. Broccoli is best if undercooked, because the more green that is left in broccoli, the more chlorophyll will be left to counteract the sulfur compounds that form gas.

THERAPEUTIC VALUE

All of the foods in the cabbage family, including broccoli, are best if eaten with proteins, because the combination helps drive amino acids to the brain. Broccoli is high in vitamins A and C, and is low in calories. It is beneficial to the eliminative system.

NUTRIENTS IN ONE POUND

Calories	103	Iron	5.6 mg
Protein	9.1 g	Vitamin A	9,700 I.U.
Fat	0.6 g	Thiamine	0.26 mg
Carbohydrates	15.2 g	Riboflavin	0.59 mg
Calcium	360 mg	Niacin	2.5 mg
Phosphorus	211 mg	Ascorbic acid	327 mg

BRUSSELS SPROUTS

Brussels sprouts are said to be native to Brussels, Belgium. They were cultivated in England early in the nineteenth century. Brussels sprouts were not extensively cultivated in this country until the early twentieth century, and were first grown in the delta region of Louisiana.

Brussels sprouts are a member of the cabbage family. The plant produces a number of very small heads along the stem. They are grown for the fresh market, frozen, and canned. Fresh sprouts may be steamed or boiled, using very little water. California and New York produce the greatest number of Brussels sprouts.

THERAPEUTIC VALUE

Brussels sprouts often produce gas, but some people can eat them without this effect if they are steamed or boiled over low heat. The sulfur in Brussels sprouts is needed for circulation, and they are good in the winter to help keep us warm.

NUTRIENTS IN ONE POUND

Calories	213	Iron	5.9 mg
Protein	20.0 g	Vitamin A	1,816 I.U.
Fat	2.3 g	Thiamine	0.36 mg
Carbohydrates	40.8 g	Riboflavin	0.73 mg
Calcium	154 mg	Niacin	3.2 mg
Phosphorus	354 mg	Ascorbic acid	431 mg

CABBAGE

Cabbage is a biennial herb of the mustard family and is native to the rocky shores of Europe. It can be considered one of the most ancient of the more common vegetables—there is evidence that it has been in cultivation for more than 4,000 years. Centuries of cultivation have produced other forms of the cabbage family: kale, kohlrabi, cauliflower, broccoli, and Brussels sprouts.

There are many kinds of cabbage: red, white, and green; with smooth or crinkled leaves; and with round, flattened, oblong, or conical heads. The most widely preferred type is the conical shape, or Wakefield cabbage. Other varieties include round or Savoy, flat-headed or Dutch, celery or Chinese cabbage.

This nonstarchy vegetable now furnishes one of our most important vegetable crops and is available all year. Its peak of abundance is between October and May. The largest cabbage-producing states are California, Colorado, Florida, New York, Pennsylvania, Texas, and Wisconsin.

When selecting cabbage, look for heads that are of a reasonable size and compactness, with leaves that are tender, not withered or puffy. Make sure the heads are free from soft rot, seed stems, and damage caused in shipping or freezing, by disease from insects, or by mechanical means. Cabbage should be used as soon after cutting as possible, as exposure to the air causes loss of vitamin C.

When cooking cabbage, use a stainless steel pan, and cook it over low heat. Because cabbage is a sulfur food, it is gas-forming, and this method of cooking will help avoid this. Covering the pan with a vacuum top will also help, and the mineral balance can be maintained in this type of vessel.

Cabbage should be cooked quickly, sliced or shredded to avoid longer cooking time, and served immediately. It is best combined with other vegetables, a starch, or a protein. (Apples and cabbage together are delicious.)

Because cabbage is a sulfur food, it can cause intestinal distress. It also contains a great deal of roughage, and some people find that they cannot eat raw cabbage.

THERAPEUTIC VALUE

Cabbage, both red and green, is one of the least expensive of the vitamin-protective foods, and is one of the most healthful vegetables. It is an excellent source of vitamin C. Raw cabbage juice may be taken when citrus fruits are prohibited, and can be made more palatable by combining it with a milder juice, such as celery or tomato.

Raw cabbage is a fair source of vitamin A, a good source of vitamin B_1, and contains some vitamin G. Cabbage is alkaline in reaction, high in cellulose or roughage, and has a very low calorie content. Cooked red cabbage is superior to white or green in that some people seem to be less sensitive to it. Cooked cabbage still retains a fair amount of vitamin A. The outside leaves of cabbage—those leaves that are very green—have as much as 40 percent more calcium than the inside leaves.

Cabbage contains many minerals: it is rich in calcium and potassium, and contains chlorine, iodine, phosphorus, sodium, and sulfur. Red cabbage has more calcium but slightly less of the other minerals than white or green cabbage.

When we were traveling in Switzerland, we noted how a variety of curly cabbage was used as a pack for eczema and for various leg conditions such as varicose veins and leg ulcers. This external pack was made by chopping the cabbage into fine pieces and mixing it with distilled water. The pack was placed on the affected area and wrapped with a linen cloth. The sulfur in cabbage helps destroy the ferments in the blood, and is especially good for any skin trouble when used both internally and externally. Sulfur is one of the elements that increases body heat, so people with cold feet might want to include cabbage in their diet.

Dr. Garnett Cheney, clinical professor at the University of California, has found that the use of raw cabbage juice showed promising results in the cure of ulcers of the stomach. Dr. Cheney thinks that the vitamin U in cabbage is what helps. He discovered that after his ulcer patients drank raw cabbage juice over a period of four or five days, most of their symptoms would disappear.

I do not believe, however, that raw vegetable juice in itself, no matter what kind, would give relief to most ulcer conditions. I use what we call cabbage juice diets because there are factors in cabbage juice that may accelerate healing and overcome ulcers of the stomach more readily than other juices.

Cabbage is very effective in helping overcome constipation, and sauerkraut, or sauerkraut juice, is particularly good for a sluggish intestinal tract, and for more serious cases of constipation. Sauerkraut juice, with a little lemon juice added, may be helpful for diabetes. Raw sauerkraut juice stimulates the body in general, and when mixed with tomato juice, makes a wonderful laxative. It is very high in vitamin C and lactic acid.

While in New Zealand, I spoke with the head food buyer for the Army, who told me that cabbage was considered one of the best foods for keeping a clean, clear complexion, and that servicewomen insisted on having cabbage salads to keep their complexions clear.

The Chinese variety of cabbage has a very high sodium content, as much as 35 percent. It does not produce as much gas as red and white cabbage do, as its sulfur content is only 2 percent, compared to 10 percent in red and white. Red and white cabbage contain nearly the same vitamins and minerals and are used the same therapeutically.

NUTRIENTS IN ONE POUND

Calories	95	Iron	1.7 mg
Protein	4.6 g	Vitamin A	530 I.U.
Fat	0.7 g	Thiamine	0.23 mg
Carbohydrates	17.5 g	Riboflavin	0.21 mg
Calcium	152 mg	Niacin	0.9 mg
Phosphorus	103 mg	Ascorbic acid	173 mg

CARROT

The carrot has been native to Europe since ancient times, and was introduced to the United States during the period of early colonization. Carrots soon became a staple garden crop. Today, they are one of the major truck and garden vegetables.

Depending on the variety, carrots grow to maturity and are ready for market within 70 to 120 days. They are always in season, and are produced in nearly all states. The largest carrot producers are Texas, Florida, and New York. Carrots are so easy to raise that a garden in your back yard can yield carrots that are rich in vitamins and high in mineral content.

When purchasing carrots, look for firm, smooth, well-shaped carrots of good color and fresh appearance. The tops should be fresh and green, unless they have been damaged in transit from grower to market. Carrots with excessively thick masses of leaf stems at the point of attachment are usually undesirable because they have large cores and may be woody. Look for carrots with "eye appeal."

Carrots may be utilized in the diet in many ways. The best way is to eat them raw and as fresh as possible. Raw carrot sticks and curls are attractive garnishes and appetizers. Grated carrot, steamed in a stainless steel kettle or baked in the oven and served with parsley and butter, is a nice dish. The bright color of carrots makes them appealing and appetizing to serve with dinner, in salads, with other vegetables, or with cottage cheese or apples and nuts.

Carrot tops are full of potassium, but because of this they are so bitter that the average person does not enjoy them. However, a small portion of the tops may be cut fine and put into mixed salads, or a bunch may be tied with string and cooked in broths or soups for flavoring and for their high mineral content. Lift them out before serving.

THERAPEUTIC VALUE

Because the carrot is so high in vitamin A, it has been used extensively in the diet to improve the eyesight. Carrots were used in World War II in aerial training schools to improve the eyesight of the students.

Many children have lower jaws that are underdeveloped. This deformity is usually the result of calcium deficiency in the child's early growth. Babies do not always get enough calcium, and some do not have enough raw food or other chewing foods that help promote normal growth of bones and teeth. It is good for a child to have a raw carrot with each meal. I have seen the teeth of children straighten out and the lower jaw develop in a year, when they were given a carrot to chew on before each meal.

Carrots contain a great deal of roughage. They will help in all cases of constipation.

Used as a general body builder, carrot juice is excellent. This juice is presently used in cases of severe illness, and as a foundation in cancer diets. It is delicious and nutritious when combined with other juices such as parsley, celery, watercress, endive, or romaine lettuce.

It is my belief that every home should have a juice machine. Everyone can benefit from drinking fresh vegetable juice, and carrot juice is one of the best. Some juice vendors believe that the short, stubby carrot is the most flavorful and colorful, and contains more vitamins and minerals. However, the long, slender carrot can be high in these values, too, and is also used.

NUTRIENTS IN ONE POUND

Calories	179	Iron	3.2 mg
Protein	4.8 g	Vitamin A	48,000 I.U.
Fat	1.2 g	Thiamine	0.27 mg
Carbohydrates	37.2 g	Riboflavin	0.26 mg
Calcium	156 mg	Niacin	2 mg
Phosphorus	148 mg	Ascorbic acid	24 mg

CAULIFLOWER

Cauliflower is a member of the cabbage family. The word "cauliflower" means "cabbage flower," and centuries of cultivation were necessary to produce a tight head of clustered flower buds in place of the compact leaves of the cabbage head. Although there are thirty-five or more varieties of this vegetable, there are probably not more than six or seven distinct varieties used.

Cauliflower contains sulfur compounds that easily break up and produce hydrogen sulfide, which has an offensive odor. If cauliflower is cooked too long, it will bring about the decomposition of these sulfur compounds.

Cauliflower is available all year, but the peak months are November through March. California is, by far, the largest grower; Arizona is second; and Colorado and New York are third.

The size of the vegetable has little to do with its quality. Fine quality cauliflower is creamy-white or white, clean, heavy, firm, and compact, with outer leaves that are fresh and green. Avoid cauliflower that has the appearance of being rice-like or granular, speckled, or spotted. A head that is no longer fresh may have yellowing leaves. If the leaves drop from the stalk, it is definitely not fresh.

THERAPEUTIC VALUE

The greatest amount of calcium in cauliflower is found in the greens that are around the head. Most people throw these away, but they are good when cooked with the cauliflower or cut up in salads. It is best to undercook this vegetable.

Cauliflower is easier for diabetic people to eat than cabbage. It is also good for reducing diets, because it is so low in calories, but keep in mind that its high phosphorus content means it is gas-forming.

NUTRIENTS IN ONE POUND

Calories	63	Iron	2.2 mg
Protein	4.9 g	Vitamin A	200 I.U.
Fat	0.4 g	Thiamine	0.21 mg
Carbohydrates	10 g	Riboflavin	0.22 mg
Calcium	45 mg	Niacin	1.2 mg
Phosphorus	147 mg	Ascorbic acid	141 mg

CELERY

Celery has long been native to marshy regions extending from Sweden southward to Algeria, Egypt, and Ethiopia. Ancient Oriental people gathered wild celery and brewed it as a medicinal herb for stomach maladies and for a general tonic. Wild celery has a bitter flavor and pungent odor. The early physicians seemed to think that the worse a concoction tasted, the better it was for the patient. The ancient Greeks valued it highly, and awarded celery as a prize to winners in many of their sport contests.

There is mention of a cultivated variety of celery grown in France in 1623, and in 1776 celery seed was sold in England for the growing of plants to be used in flavoring soups and stews. Celery has been grown commercially in the United States since about 1880.

Celery belongs to the same plant family as carrots, parsley, fennel, caraway, and anise. The characteristic flavor of these plants is from the volatile oils found in the stems, leaves, and seeds.

California and Florida are the two leading celery-producing states, but celery is also grown in many other states in the eastern and western United States. Celery is available all year, but its peak season is November through May. Study the market in your state and plan to use celery in abundance during the months when celery is in season.

The most desirable celery is of medium length, thickness, and solidity. The stalks should be brittle enough to snap easily. Pithy or stringy celery is not good to eat and probably has less vitamin and mineral content.

The pithiness of a celery stalk can be detected by pressing or twisting the stalk, and stringiness can be detected by breaking the stalk. Celery that has formed a seed stem probably has a poor flavor and may be bitter.

Celery is highly perishable, and should be kept refrigerated. To prepare for eating, scrub and wash thoroughly to be sure all poisonous sprays are removed. Before the tops of celery are used, they should be separated, and washed several times. If you are cooking celery tops, douse them in water that is slightly warm to insure a thorough washing.

If you are cooking celery, steam it only long enough to break down the fibers, or cook it a few minutes in a vessel with a tight lid. Use very little water. Cooked celery takes only about three hours to digest. Celery is also delicious in soup and as a seasoning in almost all cooked food.

THERAPEUTIC VALUE

Celery is fairly high in roughage and low in calories. Its high water content makes it an especially good food to eat with foods that are more concentrated, particularly heavy starches. It is an alkaline food and should be classified as a protective food. The greener stalks of celery are an especially good source of vitamin A, and celery is also a good source of vitamins B_1 and G. It is rich in chlorine, sodium, potassium, and magnesium.

As an all-around maintainer of good health, celery juice gets top billing. It is good by itself or mixed with other vegetable juices, and goes best with carrot, carrot and parsley, or apple. Celery can be juiced with fruits, vegetables, or nuts for a complete, easily digested meal.

Celery is generally known as a sodium food, and sodium is what we call the youth maintainer in the body. Sodium helps keep us young and active, and the muscles limber and pliable. Whenever there is a stiffness in the joints and creaking or cracking in the knees, we know we are lacking in sodium. Sodium is the one element that most people lack.

When the tissues, joints, and arteries get hard, there is too much calcium in the body, and a softer element is needed. The element that counteracts calcium best is sodium. It helps keep calcium in solution.

Celery should be eaten often because it is one of the best foods for keeping the body well. It neutralizes acids and is a good blood cleanser. It has protective properties that are beneficial to both the brain and the nervous system. Celery is an excellent food for people suffering from arthritis, neuritis, and rheumatism. It can help to clear up high blood pressure.

Sodium is one of the chemical elements needed so much in the walls of the stomach and in the intestinal tract. Celery is particularly good for these parts of the body. However, many times celery can be very irritating to a sensitive stomach because it contains a great deal of fiber. If irritation results, celery juice should be substituted. It is also best to avoid using raw celery leaves if there is any stomach irritation. Broths made of celery leaves, with other vegetables and milk or cream added, are good to take for stomach disturbance. The milk or cream has a wonderful soothing effect on the stomach, especially when there is excessive acidity. A broth made with celery and other vegetables is also good in an elimination diet.

Celery aids digestion, counteracts acidosis, halts fermentation, and purifies the bloodstream. Celery juice can be handled and tolerated by most people, especially children. However, many people prefer diluted celery juice, and it is very good when combined with pineapple or apple juice. Apple and celery juice combined is great for neutralizing the rheumatic acids in the body. Combine celery, parsley, and asparagus juice for kidney disorders; celery and papaya juice for asthma; celery and grapefruit juice with a pinch of pure cream of tartar for colds or sinus troubles; celery and parsley juice for fevers, gout, or arthritis; and, if the condition of the teeth is poor, combine beet greens, parsley, celery juice, and green kale. It is a nonstarchy vegetable.

Celery is best eaten raw, preferably in the form of combination vegetable salads. Use it as a balance in high protein salads such as chicken, tuna, or shrimp. Celery is particularly flavorful when cooked with tomatoes or green peppers. Its pot liquor is especially good as a base in soups and sauces.

The leaves of celery are rich in potassium, sodium, and sulfur. The raw leaves or tops are excellent in the treatment of diabetes. Because they are so tough, they should be chopped, liquefied, and added to other vegetables to lessen their strong taste. When eaten raw, the leaves are beneficial to the nerves and disorders resulting from nervous conditions. Celery leaves are also good for all acid conditions of the body.

NUTRIENTS IN ONE POUND
(one pound of celery contains 93 percent water)

Calories	218	Iron	2.7 mg
Protein	1.8 g	Vitamin A	182 I.U.
Fat	3.18 g	Thiamine	.13 mg
Carbohydrates	51.4 g	Riboflavin	.09 mg
Calcium	63.5 mg	Niacin	0.45 mg
Phosphorus	50 mg	Ascorbic acid	55 mg

CELERY ROOT (CELERIAC)

Celeriac is probably more commonly known as celery root. It is a turnip-rooted vegetable, and the root forms a solid knob just below the soil surface.

Italian and Swiss botanists gave the first description of celeriac about 1600. It became popular in Europe in the eighteenth century, but has never been popular in England or the United States.

Most of the United States' supply of celery root is grown in California. It is available from September to April.

When selecting this vegetable, choose roots that are firm. Press the tops of the roots to check for internal rot.

Celery root is usually boiled and eaten alone or in salads. To store, wrap it in plastic, and keep it very cool.

THERAPEUTIC VALUE

Celery root is high in phosphorus and potassium. It is beneficial to the lymphatic, nervous, and urinary systems.

NUTRIENTS IN ONE POUND

Calories	156	Iron	2.3 mg
Protein	7.0 g	Vitamin A	————
Fat	1.2 g	Thiamine	.20 mg
Carbohydrates	33.2 g	Riboflavin	.11 mg
Calcium	168 mg	Niacin	1.2 mg
Phosphorus	449 mg	Ascorbic acid	30 mg

CHERRY

Garden cherries originated chiefly from two species, the sour cherry and the sweet cherry. Both are native to Eastern Europe and Western Asia, where they have been cultivated since ancient times. Cherry pits have been found in prehistoric cave dwellings.

Cherries are grown in every state. Leading cherry producers are New York, Pennsylvania, Ohio, Michigan, Wisconsin, Montana, Idaho, Colorado, Utah, Washington, Oregon, and California. Washington, Oregon, and California lead in sweet cherry production, while Michigan leads in the production of sour cherries.

The Tartarian variety, which is mahogany to black in color, and medium to large in size, is a popular early-to-mid-season variety of sweet cherry. The cherry in heaviest demand for the fresh market is the Bing: an extra large, heart-shaped, deep maroon to black fruit. It is firm, high-flavored, and stands up well. Bings are on the market through the months of June and July. The Black Republican and Lambert are similar in appearance to the Bing. The Royal Ann is the leading light-colored cherry, and is used primarily for canning. It is large, is light amber to yellow with a red blush, and has a delightful flavor. The Schmidt is a dark red to black sweet cherry grown widely. The Windsor is another popular sweet cherry, and its color is dark red to almost black.

The leading sour varieties of the cherry are the Early Richmond of the East and Middle West, the Montmorenci, and the English Morello.

THERAPEUTIC VALUE

The cherry is high in iron, and is an excellent laxative as well as a wonderful blood builder. I believe that the black cherry is the best for eating.

Cherries mix well with other fruits and with proteins, but never with starches. They are wonderful in an elimination diet. The cherry should not often be mixed with dairy foods. This fruit, which has a high alkaline content, also helps get rid of toxic waste, and it has a wonderful effect on the glandular system.

Black cherry juice is wonderful for flavoring teas so that sugar can be avoided. It is a wonderful gall bladder and liver cleanser because of its high iron content. Take a six-ounce glass of black cherry juice each morning before breakfast for the gall bladder and liver.

NUTRIENTS IN ONE POUND

Calories	286	Iron	1.6 mg
Protein	5.3 g	Vitamin A	450 I.U.
Fat	1.2 g	Thiamine	.20 mg
Carbohydrates	71 g	Riboflavin	.24 mg
Calcium	90 mg	Niacin	1.7 mg
Phosphorus	78 mg	Ascorbic acid	41 mg

CHICORY

Chicory is closely related to endive. There are many varieties of chicory. They include green chicory, which is leafy; Belgian endive (also called witloof), which is a root chicory; and radicchio, also a root chicory, which is red and white. Chicory is best when tossed in a salad with other vegetables.

Green chicory is cultivated primarily in Europe, although varieties grow wild in Europe, Africa, Asia, and the United States. Belgian endive is mainly cultivated in Belgium and is prized for its delicate flavor. Radicchio is native to Italy and primarily grows there.

Radicchio is often sold with the root attached. If possible, the root should be eaten, because it is very good.

When selecting chicory, look for a fresh, crisp, green vegetable. Belgian endive, which looks like a tightly wrapped stalk, should be white or near-white. Radicchio should be crisp and fresh.

THERAPEUTIC VALUE

Chicory is an alkaline food that is good in elimination diets. It is high in vitamin C.

Tea made from chicory roots and used as an enema is a wonderful remedy for increasing peristaltic action and getting the liver to work.

NUTRIENTS IN ONE POUND (greens only)

Calories	74	Iron	3.3 mg
Protein	6.7 g	Vitamin A	14,880 I.U.
Fat	1.1 g	Thiamine	.22 mg
Carbohydrates	14.1 g	Riboflavin	.37 mg
Calcium	320 mg	Niacin	1.9 mg
Phosphorus	149 mg	Ascorbic acid	____

COCONUT

The coconut probably had its beginning in the Malay Archipelago and the tropical areas of the Americas. For over 3,000 years the coconut has been cultivated in Southern Asia and the East Indian islands. Now, the coconut palm may be found all along the tropical coasts, and its fruit is used as the principal food on many of the Pacific islands. It has been estimated that over 300,000 people use coconuts in some form every day, and the old saying: "He who plants a coconut tree plants vessels and clothing, food and drink, a habitation for himself and a heritage for his children," still holds true in many of the places where coconuts are grown today. The coconut tree reaches its maturity at seven years and produces its fruit for as many as seventy or eighty years.

The coconuts sold in our markets today are imported mostly from Honduras, Panama, and the Dominican Republic. There is some production on the Florida coast, but most of the nuts produced there are sold on the local market to winter tourists. One coconut is imported for each eleven persons in the United States, so you can see that it is one of the minor items in the produce business.

Coconuts may be found on the market all year, but October, November, and December are the peak months. A quality nut is one that is heavy for its size. When shaken, the liquid inside will slosh around. Do not choose a coconut without liquid, as this indicates spoilage, and nuts with moldy or wet "eyes" are unsound.

To crack a coconut, pierce the soft spots, or "eyes," at the top of the shell with an ice pick or other sharp object. Drain the liquid, then tap all around the hard shell with a hammer until the shell cracks and falls away. Or, after draining off the liquid, heat the coconut in the oven at 350° F for thirty minutes, and the shell will easily break away.

THERAPEUTIC VALUE

Coconuts are 70 percent fat. They should be eaten flaked, as a topping in salads. Coconut may also be used in nut butter, or liquefied, to get the milk out of it. It may also be used in combination with other vegetable and fruit juices.

Coconut milk compares to mother's milk in its chemical balance. It is quite a complete protein food when taken in its natural form. Coconut milk is made from the nut itself, by liquefying the meat. The water we get when we open up the coconut is not the milk and does not have a very high mineral content.

NUTRIENTS IN ONE POUND (meat only)

Calories	1,569	Iron	7.7 mg
Protein	15.9 g	Vitamin A	0
Fat	160.1 g	Thiamine	.24 mg
Carbohydrates	42.6 g	Riboflavin	.10 mg
Calcium	59 mg	Niacin	2.4 mg
Phosphorus	431 mg	Ascorbic acid	14 mg

COLLARD

The collard, with its close relative, kale, is one of the oldest members of the cabbage family. It is native to the eastern Mediterranean countries or to Asia Minor. It has been under cultivation for so long and has been so shifted about by prehistoric traders and migrating tribes that it is not certain which of these regions is the home of the species. Wild cabbage, from which the collard and the more highly developed horticultural forms arose, is still found growing along the coastal regions of Europe and Northern Africa. Its use by man as food antedates written history, and it is believed to have been in common use for more than 4,000 years. All the principal forms of collards known today have been cultivated for at least 2,000 years. Well before the Christian era the Greeks and Romans grew this plant. ''Coles'' (collards and kales) were described by European writers in the first, third, fourth, and thirteenth centuries.

It seems probable that the Celts may have introduced coles to France and Britain. They invaded Mediterranean lands repeatedly from about 600 B.C. and reached into the British Isles in the fourth century B.C. The English name is a corruption of the Anglo-Saxon ''coleworts'' or ''colewyrts,'' meaning literally ''cabbage plants.''

The first known mention of collards in America was in 1669, but because of their popularity in European gardens, it is probable that they were introduced somewhat earlier.

There are several varieties of collards, each of which is prepared in many ways. In recent years some finely chopped or sieved collards have been canned for baby foods, or for persons requiring a special diet. They can be used either cooked, or raw in salads, very much the same as cabbage.

THERAPEUTIC VALUE

Collards are very rich in calcium and a good source of vitamins A and C. Collards are good for the respiratory system, the digestive system, the skeletal system, the lymphatic system, the eliminative and urinary system, and the nervous system. This vegetable is invaluable to nearly every part of the body!

NUTRIENTS IN ONE POUND

Calories	82	Iron	3.3 mg
Protein	8.0 g	Vitamin A	14,020 I.U.
Fat	1.2 g	Thiamine	0.22 mg
Carbohydrates	14.7 g	Riboflavin	0.56 mg
Calcium	508 mg	Niacin	4.1 mg
Phosphorus	118 mg	Ascorbic acid	203 mg

CORN, SWEET

Corn is first recorded as having been found in North America in 1006, by Karlsefne, at a place called Hop, in the vicinity of the Taunton River. Indian corn was known to be cultivated in both North and South America, from Canada to Patagonia, long before Columbus discovered America. In 1492, he described corn as "a kind of grain called maize of which was made a very well-tasting flour." In the 1540 invasion by DeSoto, corn was found in Florida, Alabama, and Mississippi. According to research by Dr. Edgar Anderson, vast quantities of corn were found in excavations in southern Peru and northern Chile. Jars of kernels were found, as well as tassels, stalks, and leaves. In southern Mexico, water bowls and funerary urns used by the prehistoric Zapotecs were found decorated with ears of corn evidently cast from the original ears.

The Incas of Peru, the Mayans of Central America, and the Aztecs of Mexico used maize not only as a food, but as currency, fuel, smoking silk, jewelry, and building material. It was an important contribution to art in decorating temples, homes, ceramics, and toys. There are probably as many Indian legends based upon corn as there are Indian tribes. It played an important part in their festive and religious ceremonies. Quinche, a variety of corn still grown today, is said to have originated as an Incan corn from the Andean highlands, and was handed down for centuries both as a food for human consumption and for cattle feeding. Indian corn, or maize, was spread throughout the Orient by the early Spanish and Portuguese travelers and may have crossed the Pacific in pre-Columbian times.

Sweet corn probably originated with the North American Indians. The first written description of it is dated 1801. It is described as "having a white, shriveled grain when ripe, as yielding richer juice in the stalks than common corn.'' After sweet corn was introduced to Plymouth, it gradually became known as a common garden vegetable, and some thirty varieties were listed in the early seed catalogs of 1880.

In 1940, a vast number of varieties of sweet corn were being grown for the fresh market. This was because new hybrids suitable for cultivation in the southern and the western United States were being developed.

The most important varieties of sweet corn grown commercially are the yellow hybrids. They are more desirable for their high quality and superior food value than the white hybrids.

In the last three or four years the market season for sweet corn has developed to year-round output. Florida and California, particularly, supply the winter market. The peak months, however, are still July through September. The frozen market has also increased the winter supply.

Good quality sweet corn has cobs that are well filled with plump, milky, bright kernels just firm enough to resist a slight finger pressure. The kernels should be filled with a thick white liquid if rich-bodied flavor is desired. If the kernels are only semi-solid or doughlike, there is little sweetness and the kernel skins will be tough. The husks should be fresh and green. Yellowed husks indicate age or damage. Quality can best be determined by pulling back the husks and examining the kernels. Note, when buying, whether the corn is sweet corn or the green field corn variety. Choose the fresh, yellow corn for greater nutrition.

THERAPEUTIC VALUE

Corn is considered one of the easiest foods to digest. It is very high in roughage, so if you are following a soft diet, you should avoid it.

Corn is rated among brown rice and barley as one of the best balanced starches. For those who want to avoid weight gain, corn should be used sparingly, because it is rich in carbohydrates.

We believe that yellow corn is the best corn to use, as it is very high in magnesium, which is a wonderful bowel regulator and one of the chemical elements we need so much. Southern yellow corn is a greater bone and muscle builder than northern white corn. Yellow corn is higher in phosphorus than white corn, which makes it an excellent food for the brain and nervous system.

A yellow corn broth, or gruel, is quite soothing to the intestinal tract and, mixed with barley or brown rice, has a wonderful flavor.

Yellow corn, or yellow corn meal, should be used at least once a week in a balanced diet.

NUTRIENTS IN ONE POUND

Calories	297	Iron	1.6 mg
Protein	11.9 g	Vitamin A	1,260 I.U.
Fat	3.9 g	Thiamine	0.48 mg
Carbohydrates	66.0 g	Riboflavin	0.37 mg
Calcium	29 mg	Niacin	5.4 mg
Phosphorus	386 mg	Ascorbic acid	30 mg

CRANBERRY

Cranberries are native to the swampy regions of both the temperate and arctic zones of North America and Europe. Because they grow on slender, curved stalks, suggesting the neck of a crane, they were named "crane-berry," or "cranberry."

Long before the first colonists arrived in this country the cranberry was in common use by the Indians. The Pilgrims found them in the low marshes near the shore on the Cape Cod peninsula, and the women preserved them as a delicacy and served them with wild turkey at Thanksgiving and Christmas feasts.

Cultivation of the cranberry began early in the nineteenth century. The earliest records show that the business was largely carried on by retired seamen. Howe and McFarlin were the names of two of these men, and important varieties of cranberries are named for them. By 1870, a flourishing business had developed. It was recorded in 1832 that "Captain Henry Hall of Barnstable, Massachusetts, had then cultivated the cranberry for twenty years," and that "Mr. F. A. Hayden of Lincoln, Massachusetts, gathered from his farm in 1830, 400 bushels of cranberries which brought him in the Boston market $600."

It has been said that the old clipper ships out of Gloucester, New Bedford, and the "Down East" ports carried supplies of raw cranberries in casks so that the sailors could help themselves. They did this to prevent scurvy, just as the sailors of England and Southern Europe used limes to prevent this disease.

Cranberries grow on low, thick vines in a bog. The bogs are built on peat swamps that have been cleared, drained, and leveled. Water must be available and arranged so that the bog can be drained or flooded at the appropriate time. The surface, usually sand, on top of a subsoil that will hold moisture, must be level so the bog can be covered with water to a uniform depth when necessary. A cranberry bog takes three to five years to come into full production.

There are only five states that produce the greater supply of cranberries for market. They are, in order of production: Massachusetts, Wisconsin, New Jersey, Washington, and Oregon. The berries are marketed from September through March, and the peak months are October, November, and December.

The quality of the berry is determined by its roundness and size, and from its color, which varies from light to dark crimson, depending on the degree of maturity. Some varieties of cranberries are more olive-shaped or oblong. They have a fresh, plump appearance combined with a high luster and firmness. Avoid a shriveled, dull, soft-appearing berry.

THERAPEUTIC VALUE

Cranberries have a heavy acid content, and therefore should not be eaten too frequently. They increase the acidity of the urine. Because of their extremely tart taste, people drown them in sugar syrup, which makes them unfit for human consumption. They are best if cooked first; then add raisins and a little honey.

One of the finest therapeutic uses for cranberries is as a remedy for rectal disturbances, piles, hemorrhoids, and inflammation of the rectal pouch. Place slightly cooked cranberries in the rectum after each movement.

NUTRIENTS IN ONE POUND

Calories	218	Iron	2.7 mg
Protein	1.8 g	Vitamin A	182 I.U.
Fat	3.18 g	Thiamine	.13 mg
Carbohydrates	51.4 g	Riboflavin	.09 mg
Calcium	63.5 mg	Niacin	0.45 mg
Phosphorus	50 mg	Ascorbic acid	55 mg

CUCUMBER

The cucumber is said to be native to India, although plant explorers have never been able to discover a wild prototype. Cucumbers have been cultivated for thousands of years, and records indicate that they were used as food in ancient Egypt, and were a popular vegetable with the Greeks and Romans. The cucumber is one of the few vegetables mentioned in the Bible.

In 200 B.C. a Chinese ambassador traveled as far as Persia, where he saw cucumbers for the first time. Later, he brought them to China. At a later date, an English sea captain, returning from the West Indies, brought back pickled gherkins to Mrs. Samuel Pepys. Shortly after this period, cucumbers were grown in England.

Occasionally, in a collection of old glass, a plain glass tube or cylinder resembling a lamp chimney with parallel sides will turn up. This may be an English cucumber glass, a device used at one time to make cucumbers grow straight. George Stephenson, inventor of the locomotive, is credited with its invention.

Florida is the principal producer of cucumbers, supplying almost one-third of the total United States commercial crop for market. California, North and South Carolina, New Jersey, and New York are also large producers.

Cucumbers for slicing should be firm, fresh, bright, well-shaped, and of good medium or dark green color. The flesh should be firm and the seeds immature. Withered or shriveled cucumbers should be avoided. Their flesh is generally tough or rubbery and somewhat bitter. Overmaturity is indicated by a generally overgrown, puffy appearance. The color of overmature cucumbers is generally dull and not infrequently yellowed, the flesh is tough, the seeds hard, and the flesh in the seed cavity almost jelly-like. Cucumbers in this condition should not be used for slicing. Some varieties are of solid green color when mature enough for slicing, but usually a little whitish color will be found at the tip, with a tendency to extend in lines along the seams, where they advance from pale green to white, and finally yellow with age.

THERAPEUTIC VALUE

Cucumbers are alkaline, nonstarchy vegetables. They are a cooling food, especially when used in vegetable juices. Long ago it was believed that people would die from eating the peelings, but this is not true.

Cucumbers are wonderful as a digestive aid, and have a purifying effect on the bowel. It is not necessary to soak them in salt water. Serve them thinly sliced, raw, in sour cream, lemon juice, or yogurt for a delightful summer dish. They have a marvelous effect on the skin, and the old saying "keeping cool as a cucumber" is literally true because of its cooling effect on the blood.

NUTRIENTS IN ONE POUND (without peel)

Calories	39	Iron	1.0 mg
Protein	2.2 g	Vitamin A	0
Fat	0.3 g	Thiamine	0.11 mg
Carbohydrates	8.6 g	Riboflavin	0.14 mg
Calcium	32 mg	Niacin	0.7 mg
Phosphorus	67 mg	Ascorbic acid	27 mg

CURRANT

Cultivation of both red and black currants began in Europe in the Middle Ages. The black currant is a species that is actually closer to the gooseberry than the red currant—the currant and the gooseberry are closely related. Red currants are native to northeastern Europe.

Neither the currant nor the gooseberry will grow in the deep South or the warm areas of the western United States. Currants grow best where the weather is cold.

Black currants were used in times past for the treatment of scurvy. They are a fine source of vitamin C; red currants, however, do not contain as much of that important vitamin.

THERAPEUTIC VALUE

When dried, currants are valuable in combating anemia, since they contain iron, copper, and manganese. They have an alkaline reaction and work as a laxative in the body. Currants are important for the elimination of certain impurities.

The black currant is not as flavorful as the red currant, but is thought to be therapeutic for cases of arthritis and gout. The red currant has been recommended in cases of dysentery.

NUTRIENTS IN ONE POUND

Calories	240	Iron	4.9 mg
Protein	7.6 g	Vitamin A	1,020 I.U.
Fat	.4 g	Thiamine	.24 mg
Carbohydrates	58.2 g	Riboflavin	.22 mg
Calcium	267 mg	Niacin	1.3 mg
Phosphorus	178 mg	Ascorbic acid	889 mg

DANDELION

The dandelion is a common plant of the sunflower family. The smooth, erect, hollow flower stalk rises from the center of a rosette of leaves. As every American with a lawn knows too well, the dandelion bears a bright yellow flower that soon becomes a feathery ball of seeds which are blown far and wide by the wind. The name dandelion means ''tooth of a lion,'' referring to the irregular, jagged leaves. In the spring the dandelion is valuable to commercial beekeepers since it furnishes large quantities of nectar and pollen that are needed for feeding the young bees.

Dandelions are believed to be native to Europe and Asia but are now found in most of the temperate areas of the world. They are especially abundant as a weed in the eastern United States.

Dandelions have been used as food for centuries, but have been cultivated only within recent years. They make healthful hot greens for the daily leafy vegetable. Cook as you would spinach, adding a little more water. Dandelion greens also make a wonderful addition to the salad bowl.

Due to the variability of the plant and the freedom with which the seeds germinate, there are hundreds of varieties of dandelions. One of the main commercial varieties is the Improved Thick Leaved. This has large, dark, broad green leaves. In good soil the plants attain a spread of one-and-a-half feet. The leaves are thick and blanch readily. This type is also called American Improved. Another relatively important commercial variety is Common

French or French Large Leaf, with large, broad, partially toothed, thick, and easily blanched leaves. It is pink-ribbed.

Tender, fresh, young green leaves that are still attached to a portion of the root are likely to be succulent. When leaves are separated from the base they wilt rapidly. Seed stems are an indication of age and toughness. Age or damage is also indicated by wilted, flabby, yellow, or tough leaves. Avoid dandelions that show excessive dirt or insect damage.

THERAPEUTIC VALUE

Dandelion greens have more vitamin A than almost any other vegetable. They are also high in potassium, which makes uncultivated greens bitter to the taste.

Dandelion greens are a wonderful liver cleanser and are valuable in helping the flow of the bile. Dandelion tea is an excellent drink for helping the liver and gall bladder, and also in cases of rheumatism or gout. The greens stimulate the glands. Besides cleansing the liver, the body will be relieved of many toxic conditions that are indicated by eczema, skin rashes, etc.

NUTRIENTS IN ONE POUND

Calories	200	Iron	14.2 mg
Protein	12.3 g	Vitamin A	61,970 I.U.
Fat	3.2 g	Thiamine	.85 mg
Carbohydrates	40 g	Riboflavin	0.65 mg
Calcium	849 mg	Niacin	3.8 mg
Phosphorus	318 mg	Ascorbic acid	163 mg

DATE

The date tree was cultivated as far back as 3,500 B.C. in Mesopotamia (modern Iraq) and in the Nile Valley of Egypt. These old date trees would range in height from forty to eighty feet. Dates were known as "the candy that grows on trees."

Early Spanish missionaries introduced the date tree to the Western World, and some of the original palms or their offshoots are still found in parts of California and Mexico, where the missions were first established. It was not until the middle of the nineteenth century, when they started planting date trees in the warmer interior valleys, that the date began to show promise commercially.

Dates are sweet and tasty when eaten fresh. This fruit can be preserved by drying it and either pressing it into cakes or packing it individually after the drying process. Centuries ago, Arabian caravans relied on dates in this dried form as their principal food on long journeys across the desert.

Fresh dates also add to many dishes, such as fruit salads, cereals, muffins, cookies, and cakes. Other date products are date honey, date sugar, date sap (an intoxicating drink), and date palm flour, which is made from the pith of the tree. Fresh dates keep well under refrigeration and are therefore available throughout the year. The season of top abundance is from September to May, and the peak is in November.

Dates are classified as soft, semidry, or dry, depending upon the softness of the ripe fruit. Another classification is according to the kind of sugar contained in the ripe fruit—invert-sugar dates contain dextrose and glucose, and cane-sugar dates contain mostly cane sugar (sucrose). Most of the soft varieties are invert-sugar dates, while most dry varieties are cane-sugar dates. The dry varieties contain only a little moisture when ripe and are nonperishable, while the soft or semidry varieties contain a considerable amount of moisture and are more perishable unless dried by either natural or artificial means.

The fully ripe date is plump, with a golden-brown smooth skin. The natural sugar contained in the date is much better for a person than highly refined white sugar. Dates that are pitted, stuffed with walnuts, and rolled in coconut are a delight to anyone. Children like dates and date-nut candies, and these are great to put in their lunches. Dates are certainly better for them than ordinary candy bars.

Dates in their dry, powdered form can be used as sugar. Dates used on cereals make a wonderful sweetener. They can be used finely cut in salads to change the flavor or added to breads and baked goods to give the natural sweet taste that many like. Stuffed dates with nut centers, nut butter centers, or coconut and honey centers, are wonderful for children.

The Deglet Noor date, which is native to Algeria, is the leading commercial variety in the United States. It is grown chiefly in the Coachella Valley of California, and accounts for about 85 percent of the total date crop. The fruit is medium to rather large; ovate-oblong; coral red, ripening to amber, and curing to a deeper brown.

Dates go well with cottage cheese, or any other cheese, and with apples or any semiacid fruit. However, it is best not to mix them with very watery fruits, such as oranges, lemons, grapefruit, and watermelon.

Domestic dates are usually fresh. The only processing they receive is cleaning, pasteurizing, and either reducing or increasing the moisture as necessary for the best storing or eating quality. A so-called "cured" date is one that has either reached the proper state of dryness on the tree or has been dried after picking. The purpose of such drying is to reduce the moisture below the point at which the date would sour or mold.

THERAPEUTIC VALUE

Dates can be eaten with whole raw milk for ulcers of the stomach. When used this way, however, you may have to soak and peel them. The date water can be used with milk for children who have sensitive stomachs, as it helps digest the milk.

The fiber or cellulose of the date is very soft and will not irritate a sensitive bowel or stomach. Dates are heat-producing, and give energy to people who engage in physical exercise and hard work. They are also a good source of copper, which is a diet essential, even though it is needed by the body only in small amounts.

NUTRIENTS IN ONE POUND

Calories	1,121	Iron	24 mg
Protein	8.7 g	Vitamin A	200 I.U.
Fat	2.4 g	Thiamine	.35 mg
Carbohydrates	297.8 g	Riboflavin	.38 mg
Calcium	284 mg	Niacin	8.6 mg
Phosphorus	237 mg	Ascorbic acid	0 mg

EGGPLANT

Eggplant is an annual plant. It belongs to the potato family, and is native to India, where it has been grown for thousands of years. Eggplant has large white to dark purple fleshy fruit that can be as large as six or eight inches in diameter. The Chinese and Arabs grew eggplant as early as the ninth century, and it is said to have been introduced into Europe by the early invaders. British traders brought this vegetable to the London market from West Africa in the seventeenth century, calling it "Guinea squash."

According to available records, the early types of eggplant had small fruits of ovoid shape. This, perhaps, accounts for its name.

Eggplant is available all year. Florida, California, Texas, Louisiana, and New Jersey produce most of the eggplant in this country.

Black Beauty and Fort Myers Market are the types most extensively grown, and Improved Large Purple and New York Purple are well-known varieties. Black Beauty plants grow to a height of twenty-four to thirty inches and have fruits that are globe-shaped with a slight tendency to being triangular. The fruits of this variety are large and symmetrical, but are thicker and broader than those of other varieties, and retain their glossy black-purple coloring for a long time. Fort Myers Market grows to three feet in height, and the fruits are long, oval, and deep black.

When selecting eggplant, choose those that are heavy and firm. They should have a uniform dark color and be free from blemish.

Eggplant is best steamed or baked. Cheese and tomatoes can be added for flavoring.

THERAPEUTIC VALUE

Eggplant is low in calories and is a nonstarchy fruit that is cooked as a vegetable. It contains a large amount of water. It is good for balancing diets that are heavy in protein and starches.

NUTRIENTS IN ONE POUND

Calories	111	Iron	1.6 mg
Protein	4.3 g	Vitamin A	100 I.U.
Fat	0.8 g	Thiamine	0.27 mg
Carbohydrates	21.7 g	Riboflavin	0.22 mg
Calcium	59 mg	Niacin	3.2 mg
Phosphorus	146 mg	Ascorbic acid	19 mg

ELDERBERRY

Elderberries are small red or black berries that grow on the elder bush. The elder is a member of the honeysuckle family, and two varieties are native to the United States. One type grows in the East and the Midwest, and the other in the West. The western variety are blue and are rarely eaten; the eastern type are black, and good in pies and jellies. This fruit is rarely grown commercially. It is high in vitamin C.

Elderberries may be eaten with other fruits. They are good mixed in fruit salad.

THERAPEUTIC VALUE

Elderberry juice is wonderful as a cleanser. It is very rich and may have to be diluted with other juices. Elderberry juice is especially good as a tonic for the reproductive and glandular system, and elderberry blossoms, when dried, can be used as a kidney tea.

NUTRIENTS IN ONE POUND

Calories	307	Iron	6.8 mg
Protein	11.1 g	Vitamin A	2,560 I.U.
Fat	2.1 g	Thiamine	.30 mg
Carbohydrates	69.9 g	Riboflavin	.27 mg
Calcium	162 mg	Niacin	2.3 mg
Phosphorus	119 mg	Ascorbic acid	154 mg

ENDIVE AND ESCAROLE

Native to the East Indies, endive and escarole were introduced into Egypt and Greece at a very early period and references to them appear in history. The plants were brought to America by colonists. Endive is closely related botanically to chicory and the two names are sometimes incorrectly used as synonyms. Escarole is another name for a type of endive with broad leaves and a well-blanched heart. The word ''endive'' is used to designate plants with narrow, finely divided, curly leaves.

These greens are used raw in salad, or may be cooked like spinach. The slightly bitter flavor adds zest to a mixed salad.

There are two general types of this vegetable, the narrow leaf (endive), and broad leaf (escarole). Finely curled types include: Pancalier, with large, curly, deeply cut leaves, a blanched center, and midribs tinged with rose; Green Curled Ruffee, with deeply cut leaf margins, a medium-sized head, and midribs or stems of pure green; and White Curled, with a plant twelve to thirteen inches in diameter, leaves that are finely curled, and broad ribs that are slightly rose tinged but creamy white at the heart.

Broad-leaved types include: Full Heart Batavian, which is a medium-large plant with a very deep, full, compact, well-blanched heart of infolded broad leaves that are of thick buttery texture; and Cos Type Batavian, with leaves broader and rounder than Full Heart Batavian, and blanching to a clear yellow.

Crispness, freshness, and tenderness are essential factors of quality. Wilted plants, especially those that have brown leaves, are undesirable, as are plants with tough, coarse leaves. Such leaves will be excessively bitter. Tenderness can be determined by breaking or twisting a leaf. In the unblanched condition leaves should be green, but when blanched, center leaves should be creamy white or yellowish white.

THERAPEUTIC VALUE

Escarole and endive are very high in vitamin A, and work very well in ridding the body of infections. They are both high in iron and potassium and are alkaline in reaction. Escarole and endive are both useful as an appetite stimulant because of their bitter ingredients. Escarole also helps to activate the bile. They are best when used raw.

NUTRIENTS IN ONE POUND (both escarole and endive)

Calories	80	Iron	6.8 mg
Protein	6.8 g	Vitamin A	13,170 I.U.
Fat	.4 g	Thiamine	.27 mg
Carbohydrates	16.4 g	Riboflavin	.56 mg
Calcium	323 mg	Niacin	2.0 mg
Phosphorus	216 mg	Ascorbic acid	42 mg

FIG

The fig was used by man in ancient civilizations. Native to western Asia and the Mediterranean areas, it spread to Arabia, Syria, and Israel. The ancient Semites were known to have carried this fruit from Phoenicia to their Mediterranean colonies as early as the second century B.C.

The ancient Greeks introduced the fig to neighboring countries, although at one point in Greek history it was so highly prized that its exportation was forbidden by law. Greek writers such as Homer, Herodotus, Aristophanes, and Plato eulogized the fig in their writings.

The Romans were known to have transported figs into all the temperate zones of Europe. In many southern European countries today this fruit is thrown at newlyweds in much the same manner that rice is thrown in America.

In the middle of the eighteenth century the Spanish mission fathers introduced this fruit to California, where the fig is still one of the main fruit crops. Now it is also grown in the southeastern states and along the Gulf of Mexico. It prospers and produces huge crops in these regions.

About 87 percent of the fig crop is dried for market. Three pounds of fresh figs are required to make one pound of dried figs. The fresh fruit is usually available from June through November. The peak months are September and October.

Both white and black figs are highly perishable when fresh. To be of good quality they must be fully ripe. The ripe fig is rather soft. An overripe fig can be detected by a sour odor, which is due to fermentation of the juice. The color ranges from greenish-yellow to purplish or almost black, according to the variety of the fruit. Bruised fruit should be avoided, as decay sets in very quickly.

THERAPEUTIC VALUE

Figs have a high sugar content, so they are great producers of energy. They are best eaten raw and fresh; however, dried figs may also give nourishment to the body, especially in the winter. Figs are laxative because of the mucin and pectin they contain. They are a high calcium food, high in carbohydrates, and turn into energy very quickly.

It is always best to use unsulfured figs. The black figs are high in potassium. One remedy for arthritis is black Mission figs and raw goat milk. This combination acts as a purge to the intestinal tract, and helps develop a toxin-free body.

In fresh form, figs will mix with all fruits. Dried figs will mix well with starches, vegetables, and subacid fruits, but not with acid fruits such as tomatoes, grapefruit, and oranges.

Figs can be stuffed with cheese or nut butters and used for weight building. When used in combination with soy milk they are wonderful between meals as a weight builder. Fig juice can be used with other juices to balance their nonlaxative qualities. The fig sugars are very quickly taken up by the intestinal tract and used by the body.

NUTRIENTS IN ONE POUND

Calories	357	Iron	2.7 mg
Protein	6.4 g	Vitamin A	360 I.U.
Fat	1.8 g	Thiamine	0.25 mg
Carbohydrates	89 g	Riboflavin	0.23 mg
Calcium	245 mg	Niacin	2.5 mg
Phosphorus	145 mg	Ascorbic acid	7 mg

GARLIC

Garlic is native to Western Asia and the Mediterranean area. It has been in cultivation for centuries.

Garlic is usually closely associated with the onion, probably because they both have a pungent taste and similar appearance. Surprisingly enough, however, garlic is a member of the lily family and has a very distinguished background.

Garlic is referred to in the Bible, and it was used by the early Egyptians for both cooking and embalming purposes. Homer and Herodotus also mentioned it in their writings. The ancient Romans believed garlic possessed magical powers, and they fed it to their soldiers to make them courageous. Europeans, especially the Italians and Spanish, have used it regularly for 2,000 years and more. The Spanish are believed to have brought garlic to the New World where it became an immediate favorite with the Indians. They liked it better than any of the other root or bulb crops from Europe.

Garlic is a bulbous-rooted perennial plant. The root is a compound bulb consisting of several smaller sections or cloves which are enveloped by a common skin or membrane. A garlic bulb has a strong odor and an acrid flavor. It differs from the onion only by being more potent in its effects.

The medicinal qualities of garlic have been celebrated since earliest times. Pliny said it was a remedy for sixty-one ailments, while Aristotle, Dioscorides, and Sotion also sang its praises in their writings.

Garlic has been used as an attempted cure for dropsy, phrenitis, jaundice, scrofulous swellings of the neck, bronchitis, tuberculosis, and many other ailments.

Three important varieties of garlic are the Creole or American, the Italian or Mexican, and the Tahiti. The Creole, a white skin variety, is the strongest of the three. The Italian is noted for its many cloves and usually has a pink skin. The largest of these varieties is the Tahiti, which measures from two to three inches in diameter.

Good quality garlic is thoroughly dry with firm and well-shaped cloves. There is little difference in flavor between the white and red varieties.

THERAPEUTIC VALUE

Garlic has long been considered a medicinal plant. It is high in iodine and sulfur. The body converts garlic to alkaline ash in the process of digestion. It can be mixed with parsley and used in the treatment of high blood pressure, is a remedy for worms, and can be used as a pack on the throat for the treatment of goiter, because of its high iodine content. Garlic oil enemas are used to cleanse the bowel of pinworms. This enema is prepared by emptying the contents of two garlic oil capsules (after slitting them with a razor blade) into one pint of water.

Garlic has been used in cases of tuberculosis in a large sanitarium in Europe with great success. The patients at this sanitarium lived on garlic for a period of one to two months. It is believed that it has a favorable effect on the mucous membranes of the throat and the air passages of the lungs, and is extremely helpful in cases of asthma and hay fever.

Chopped garlic, steeped in a little peanut or soy oil, makes an excellent pack for chest colds and bronchial or pulmonary infections. Garlic is an internal antiseptic, and contains a bacteriocide called crotonaldehyde. It has a stimulating effect on the sexual glands.

NUTRIENTS IN ONE POUND

Calories	547	Iron	6.0 mg
Protein	24.8 g	Vitamin A	trace
Fat	.8 g	Thiamine	1.01 mg
Carbohydrates	123.0 g	Riboflavin	.31 mg
Calcium	116 mg	Niacin	1.9 mg
Phosphorus	806 mg	Ascorbic acid	59 mg

GOOSEBERRY

There are approximately fifty species of gooseberries in the Northern Hemisphere. The greatest number are found in North America. The gooseberry is native to Europe, Western Asia, and northern Africa.

Gooseberries were cultivated in home gardens in the Low Countries of Europe from the beginning of the sixteenth century. The English brought the European gooseberry to a high stage of cultivation. Enthusiasm for this fruit is reflected in gooseberry shows given yearly to encourage development of better varieties. In 1629 there were descriptions of three red varieties, one blue, and one green. By 1831 there was a published list of 722 varieties. It is believed that the gooseberry was brought from England in 1629 by the Massachusetts Bay Colonists. However, because of the humid eastern summers, the gooseberry is subject to mildew, and thrives best in the Pacific coast regions. Because the gooseberry plants attract a fungus that attacks the white pine, federal and state laws regulate interstate shipments of gooseberries and the areas in which they may be grown.

Some people say that the origin of the name gooseberry stems from the fact that it is often served with goose. Another belief is that it got its name from the Dutch "kruisbes," meaning "crossberry." Although the English enjoy uncooked ripe gooseberries, Americans enjoy them more in pies, tarts, jams, jellies, conserves, preserves, and marmalades. Gooseberries also are used in spiced dishes, and are often combined with other fruits. Also, the juice from gooseberries can be used alone or combined with other fruit juices.

THERAPEUTIC VALUE

Gooseberries are considered to be good for the liver and intestinal tract. They develop an alkaline ash when digested. Gooseberries are watery, and have a high potassium and sodium content. When cooking them, date sugar is best for sweetening, or honey can be added after cooking.

NUTRIENTS IN ONE POUND

Calories	178	Iron	2.3 mg
Protein	3.6 g	Vitamin A	1,330 I.U.
Fat	0.9 g	Thiamine	0 mg
Carbohydrates	44 g	Riboflavin	0 mg
Calcium	100 mg	Niacin	0 mg
Phosphorus	127 mg	Ascorbic acid	149 mg

GRAPE

The grape is one of the oldest fruits in history. Grape seeds have been found in mummy cases in Egyptian tombs that are more than 3,000 years old. At the time of Homer, the Greeks were using wines, and the Bible tells of grape cultivation in the time of Noah. North America was known to the Norse sea rovers as "Vinland" because the grapevines were so abundant.

The Mission Fathers of California were the first to grow the European type of grape. This variety became known as the Mission grape and remained the choice variety until 1860 when other choice European varieties were introduced into this country.

Between 6,000 and 8,000 varieties of grapes have been named and described, but only 40 to 50 varieties are important commercially. Table grapes must be attractive in appearance and sweet and firm. Large size, brilliant color, and beautifully formed bunches are the qualities desired.

There are four classes of grapes: wine grapes, table grapes, raisin grapes, and sweet (unfermented) juice grapes. The big grape producing states, in addition to California, are New York, Michigan, and Washington.

Domestic grapes are available from late July through March, and the peak is from August to November. Grapes are also imported from February through May from Argentina, Chile, and South Africa.

Emperor grapes are a Thanksgiving and Christmas favorite. The clusters are large, long, and well-filled. The fruit is uniform, large, elongated obovoid, light red to reddish-purple, seeded, and neutral in flavor, and the skin is tough. They are on the market in October and well into March.

Thompson Seedless were first grown in California near Yuba City by Mr. William Thompson and are now very popular. The clusters are large, long, and well-filled; the fruit uniform, medium-sized, and ellipsoidal. The color is greenish-white to light golden. They are seedless, firm, and tender, and are very sweet when fully ripened. They are moderately tender-skinned. Thompson Seedless grapes are on the market from late June into November.

The Tokay variety grows in large clusters that are conical and compact. The grapes are large, ovoid with a flattened end, and brilliant red to dark red. They are seeded, very firm, neutral in flavor and have thick skins. Tokay grapes are on the market from September into November.

Other table varieties include Almeria, Cornichon, Red and White Malaga, Ribier, Lady Fingers, Catawba, Delaware, and Niagara.

The principal juice grape is the Concord, a leading native grape, that is blue-black in color, medium-sized, and tough-skinned. It is also used as a table grape and is on the market in September and October.

THERAPEUTIC VALUE

Grapes are used throughout the world for curative purposes. In France, it is not uncommon for people to use grapes as their sole diet for many days during the grape season. The low incidence of cancer in these areas has been attributed to the high percentage of grapes in the daily diet. The therapeutic value of grapes is said to be due to high magnesium content. Magnesium is an element that is needed for good bowel movements. Grapes are wonderful for replacing this chemical element.

The juice of the Concord grape is one of the best to use. Juice from other grapes, however, can be used as well. If the juice is too sweet or upsets the stomach, a little lemon juice can be added. Mix with pineapple juice or any citrus fruit, if desired. Used in combination with whey, soy milk, and an egg yolk, it makes a wonderful tonic for the blood. When purchasing bottled grape juice, be sure it is unsweetened.

Grape skins and seeds are good for bulk, but sometimes are irritating in conditions of colitis and ulcers, so they should not be eaten by persons who have these conditions.

When chewed well, bitter grape skins make a good laxative. There is also a laxative element found in the seeds.

Grapes are wonderful for promoting action of the bowel, cleansing the liver, and aiding kidney function. They are alkalinizing to the blood, and high in water content, so they add to the fluids necessary to eliminate hardened deposits that may have settled in any part of the body. They are wonderful for the kidneys and the bladder and are very soothing to the nervous system. The high content of grape sugar gives quick energy. Dark grapes are high in iron, which makes them good blood builders.

As grapes do not mix well with other foods, it is best to eat them alone. Make sure they are ripe, as the green acids are not good for the blood. They also make a wonderful snack for children—they are sweet, and much better for them than candy!

Crushed grapes may be used as a pack on a tumor or growth. Any infected area will improve after a grape pack is applied. It can be placed on the area of disturbance for a period of three to four days.

A one-day-a-week grape diet is good, during the grape season. It can be used when elimination is desired.

NUTRIENTS IN ONE POUND

Calories	324	Iron	2.6 mg
Protein	3.5 g	Vitamin A	330 I.U.
Fat	1.8 g	Thiamine	0.24 mg
Carbohydrates	73.5 g	Riboflavin	0.12 mg
Calcium	75 mg	Niacin	1.9 mg
Phosphorus	92 mg	Ascorbic acid	17 mg

GRAPEFRUIT

The name "grapefruit" originated in the West Indies in the eighteenth century, perhaps because of the fact that its fruit grows in clusters of three to twelve or more, similar to grape clusters. This citrus fruit was cultivated more than 4,000 years ago in India and Malaysia, but it was not until the sixteenth century that it was introduced to this country by the Spaniards. For many years it was not popular because of its slightly bitter taste. From 1880 to 1885 a group of Florida grapefruit growers shipped crates of the fruit to Philadelphia and New York and encouraged people to try it. In about 1915 the commercial sale of grapefruit expanded, until its production spread into three other states—California, Arizona, and Texas.

The United States furnishes about 97 percent of the world's supply of grapefruit, and Florida and Texas together produce about 90 percent of the grapefruit grown in the United States. The Marsh seedless grapefruit is the most popular variety today.

The grapefruit tree is about the size of the orange tree and reaches a height of twenty to forty feet. Like the orange, it blooms in the spring. In California and Arizona, the fruit ripens throughout the year. Although grapefruit is available all year, it is most abundant from January through May. Grapefruit is also imported by the United States from Cuba in the late summer and early fall.

Grapefruit of good quality is firm, but springy to the touch, well-shaped, and heavy for its size—the heavier the fruit, the better. Do not choose soft, wilted, or flabby fruit. The heavy fruits are usually thin-skinned and contain more juice than those with coarse skin or those puffy or spongy to the touch.

Grapefruit often has a reddish brown color over the normal yellow, which is called "russeting." Russeting does not affect the flavor in any way. Most of the defects found on the skin of the grapefruit are minor and do not affect the eating quality of the fruit. However, fruit with decayed spots is not desirable, as the decay usually affects the flavor. Decay may appear as a soft, discolored area on the stem end of the fruit or it may appear as a colorless area that breaks easily when pressure is applied. If the skin of the fruit appears rough, ridged, or wrinkled, it is likely to be thick-skinned.

THERAPEUTIC VALUE

Grapefruit is a subtropical acid fruit, and is highly alkaline in reaction. It is best eaten with other acid fruits, nuts, or milk. Eat grapefruit immediately after cutting into the rind to benefit from all of its goodness. For best digestion and assimilation, avoid eating grapefruit with sweeter fruits or with starches. The grapefruit is less acidulous than the lemon and is a good substitute when oranges or their juice cannot be tolerated, or when the alkaline reserves in the body need to be augmented.

Grapefruit is rich in vitamins C and B_1 and is a good source of vitamin B_2. It is low in calories, which makes it a good drink on a reducing diet. There is less sugar in grapefruit than in oranges. Eat the sun-ripened fruit when possible, as this fruit needs no sweetening, and is better for you. If sweetening is necessary, use a little honey.

Grapefruit is very rich in citric acids and their salts, and in potassium and calcium. Use it often in combination with meats, because grapefruit juice is excellent as an aid in the digestion of meats. However, avoid the overuse of all citric acid fruits as they are a powerful dissolver of the catarrhal accumulations in the body and the elimination of too much toxic material all at once may cause boils, irritated nerves, diarrhea, and other problems. People are often so eager to get vitamins and minerals into the body that they sometimes do not consider that the powerful action of citric acid causes irritation and discomfort.

When taken right before bedtime, grapefruit is conducive to a sound sleep. A drink of grapefruit juice first thing in the morning helps prevent constipation. It is also an excellent aid in reducing fevers from colds and the flu, and seldom causes allergic reactions.

Grapefruit rind contains the very valuable vitamin P, which is an important vitamin for healthy gums and teeth. This vitamin may be extracted by simmering the rind in water for about twenty minutes. Strain, and drink.

The sour taste of grapefruit increases the flow of digestive juices in the stomach. Grapefruit served at the beginning of a meal stimulates the appetite and helps in digestion.

This fruit is also good for any hardening of body tissue, such as hardening of the liver and the arteries. It can also help prevent stone formations.

NUTRIENTS IN ONE POUND

Calories	133	Iron	0.9 mg
Protein	1.5 g	Vitamin A	4,770 I.U.
Fat	0.6 g	Thiamine	0.11 mg
Carbohydrates	30.3 g	Riboflavin	0.06 mg
Calcium	51 mg	Niacin	0.06 mg
Phosphorus	54 mg	Ascorbic acid	12 mg

GUAVA

The guava is called the apple of the tropics. It is native to tropical America, but has been dispersed throughout all equatorial regions. It is grown in subtropical Florida and California, and the tree is a hearty one.

The guava tree produces large quantities of fruit. The fruit is round, with a white or yellow skin and a pulp of the same color, although the pulp is sometimes crimson. It ranges from the size of a large cherry to that of a pear or apple.

THERAPEUTIC VALUE

The guava is subacid and alkaline in reaction. It has a high vitamin C content, and also contains potassium, phosphorus, sulfur, and chlorine. It is good for the skeletal and lymphatic systems.

NUTRIENTS IN ONE POUND

Calories	273	Iron	4.0 mg
Protein	3.5 g	Vitamin A	1,230 I.U.
Fat	2.6 g	Thiamine	.23 mg
Carbohydrates	66 g	Riboflavin	.21 mg
Calcium	101 mg	Niacin	5.1 mg
Phosphorus	185 mg	Ascorbic acid	1,065 mg

HUCKLEBERRY

The huckleberry resembles the blueberry, but does not belong to the blueberry family. Although all huckleberries are edible, some species are not very tasty.

The garden huckleberry, which was developed by Luther Burbank, is closely related to the tomato. It is best in pie, with lemon juice added.

When eating huckleberries, add a little honey. They can also be mixed in fruit salads.

THERAPEUTIC VALUE

Huckleberries are especially helpful in aiding the pancreas in digesting sugars and starches. This fruit is alkaline in reaction.

The huckleberry is high in vitamins B and C and potassium. They can be used in an elimination diet, and because they are high in iron, are good for building the blood.

Huckleberries have been used as packs on running sores, eczema, and skin disorders. The leaves of the huckleberry may be dried and used to make a tea that is good for poor starch digestion.

KALE

Kale, and collard, its close relative, are the oldest known members of the cabbage family. Wild cabbage, which strongly resembles kale in its appearance, is still found growing along the European coasts and in North Africa. Kale is native either to the eastern Mediterranean region or to Asia Minor. It is known that man has been eating this vegetable for more than 4,000 years.

The word "kale" was first used in Scotland, and is derived from the Greek and Latin words "coles" and "caulis." These words refer to the whole group of cabbage-like plants. In America, kale was first mentioned in 1669, although it was probably introduced to this continent at an earlier date.

The sulfur compounds that are found in the cabbage family are, of course, also found in kale. These compounds break up easily, and decomposition occurs when kale is cooked too long or at too low a temperature. Overcooking also destroys the flavor.

Kale is on the market all year, but is most abundant through the late fall and winter. The peak months are December through February. Kale comes principally from Virginia, New York, New Jersey, and the Middle Atlantic states.

There are now many varieties of kale, but the crinkly-leaved and the smooth-leaved are the two most popular commercial types. The smooth type is usually referred to as spring kale, and the curly as green Scotch kale, or Siberian blue kale. Scotch kale are usually crinkled and curled, have a finely divided leaf, and are bright green to yellowish-green in color. The leaves of the Siberian kale are flattened and smooth in the centers, with curled and ruffled edges, and are of a deep, bluish-green color. Wilted and yellowed leaves should be avoided.

THERAPEUTIC VALUE

Kale is very high in calcium, vitamin A, and iron. It is good for building up the calcium content of the body, and builds strong teeth. Kale is beneficial to the digestive and nervous systems.

NUTRIENTS IN ONE POUND

Calories	117	Iron	6.4 mg
Protein	11.3 g	Vitamin A	21,950 I.U.
Fat	1.7 g	Thiamine	0.30 mg
Carbohydrates	21.0 g	Riboflavin	0.76 mg
Calcium	655 mg	Niacin	5.8 mg
Phosphorus	180 mg	Ascorbic acid	335 mg

KOHLRABI

Kohlrabi is native to Northern Europe, and is a member of the cabbage family. The name "kohlrabi" is taken directly from the German and means "cabbage turnip." A European botanist first described this vegetable in 1554, and by the end of the sixteenth century it was known in Germany, England, Spain, Italy, Tripoli, and the eastern Mediterranean. The first mention of its presence in the United States was in about 1800. Kohlrabi is grown for the swollen stem, which resembles white turnips in flavor, but is more delicate.

Kohlrabi has an unusual appearance that distinguishes it from other members of the cabbage family. Instead of a head of closely packed leaves, there is a globular swelling of the stem, some three or four inches in diameter just above the ground. The leaves are similar to those of a turnip.

The leading types of kohlrabi in this country are the White and Purple Vienna. The White variety, which is actually a light green, is the more popular. Other varieties with fancy leaves are grown in Europe. These are used chiefly for ornamental purposes.

The condition of the tops is a good indication of quality. The tops should be young and green. The thickened stem should be firm and crisp, and not over about three inches in diameter.

THERAPEUTIC VALUE

Kohlrabi has an alkaline reaction and is high in vitamin C. Because it has a high vitamin C content, it is good for the skeletal, digestive, and lymphatic systems. It may be baked or steamed.

NUTRIENTS IN ONE POUND

Calories	73	Iron	1.5 mg
Protein	5.1 g	Vitamin A	trace
Fat	0.2 g	Thiamine	0.14 mg
Carbohydrates	16.4 g	Riboflavin	0.12 mg
Calcium	113 mg	Niacin	0.6 mg
Phosphorus	122 mg	Ascorbic acid	149 mg

LEEK

The leek, like the common onion and garlic, originated in middle Asia, with secondary centers of development and distribution in Western Asia and the Mediterranean lands. It has been cultivated for food since prehistoric times. It is the "prason" of the ancient Greeks and the "porrum" of the Romans. They distinguished two kinds—leek and chives. The leek forms a cylindrical bulb with a flattened, solid leaf, while the chive was supposedly developed by thick planting.

In Europe and the British Isles, leeks are a favorite vegetable for soups and broths because of their delicate onion flavor. Often called "the poor man's asparagus," leeks may be prepared in the ways suitable for asparagus. By removing the outside leaves, cutting the green part down to five or six inches in length, and cutting off the root, they will cook quickly—fifteen to twenty minutes. They may be served hot or cold, with milk, cream, mushroom sauce, or other sauces.

THERAPEUTIC VALUE

Leeks are best used in soups and broths, and sometimes in vegetable juices. They are good for throat disorders and acute nasal discharges, because they loosen the phlegm. Leeks are a good blood purifier and are good for the liver and the respiratory system.

NUTRIENTS IN ONE POUND
(edible portion of lower leaf and stem)

Calories	204	Iron	5 mg
Protein	10 g	Vitamin A	183 I.U.
Fat	1.4 g	Thiamine	0.5 mg
Carbohydrates	46.8 g	Riboflavin	0.3 mg
Calcium	236 mg	Niacin	2.3 mg
Phosphorus	22.7 mg	Ascorbic acid	77 mg

LEMON

Lemons, one of the most highly alkalinizing foods, are native to tropical Asia, where cultivation dates back at least 2,500 years. In the twelfth century the Arabs brought lemons to Spain and Africa. It was Christopher Columbus, according to Las Casas, the Spanish historian, who brought the seeds of lemons with him from the Canary Islands on his second voyage.

In the New World, lemons were introduced by the Spanish adventurers in Haiti, then known as Hispaniola. In the United States, Florida was the first lemon-producing area, and this state led in production of lemons until a heavy freeze in 1895 killed the lemon groves. They were never replanted. Now, about 95 percent of the lemons used in the United States and Canada are produced in southern California. The other 5 percent are grown in Italy. Italy and California together produce nearly the world's entire supply of lemons.

In 1870, a variety of lemon called the Eureka was started from Sicilian lemon seed planted in Los Angeles by C. R. Workmen. The Eureka, along with the Lisbon, are the two varieties most commonly grown commercially. The Eureka grows in prolific quantity and is early-bearing, from late spring to summer; the Lisbon tends to bear only one large crop a year, in either spring or winter. A single lemon tree has been known to produce 3,000 lemons a year. This is because lemon trees bloom and ripen fruit every month of the year. The most fruit is produced between January and May.

The best lemons have skin of an oily, fine texture and are heavy for their size. This type is more apt to be full of juice, with a minimum of seeds and waste fibers. Choose lemons of a deep yellow color for ripeness and juice. They should be firm, but not hard, to the touch. Avoid using lemons that show signs of bruises, as fruits that have been mechanically injured are more subject to mold. Decay on the fruit appears as a mold or a discolored soft area at the stem end. Shriveled or hard-skinned fruits, or those that are soft or spongy to the touch, are not desirable. They may be old, dried out, mechanically injured, or affected by a rot at the center.

Lemon juice makes a good substitute for vinegar, especially in salad dressing, and for flavorings generally. Use a little lemon juice to cut the sweetness in very sweet fruit juices and use lemons in milk or cream, or canned milk, to curdle it, or when you want to make cheese. Use lemon to soften water, and when shampooing the hair, use lemon juice in the water to make an excellent rinse.

THERAPEUTIC VALUE

The lemon is rich in alkaline elements. Fresh lemon juice is an outstanding source of vitamin C. However, much of this valuable vitamin is lost if the juice is left exposed to the air too long. Lemons are high in potassium, rich in vitamin B_1, and may be considered a good source of vitamin G. Both lemons and limes contain 5 to 6 percent citric acid as compared with oranges, which contain only 1 to 1.5 percent, or grapefruit, which contain 1 to 2 percent. The lemon is classified as an acid fruit, along with other citrus fruits, cranberries, loganberries, loquats, pineapples, pomegranates, strawberries, and tamarinds.

Lemons are ideal for getting rid of toxic materials in the body, but the citric acid in lemons can really stir up the inactive acids and inactive toxic settlements of the body. The mineral content of the lemon is alkaline-forming in its ash. However, before this alkaline ash goes into the tissues, the citric acid is stirring up many of the acids in the body and it is difficult to get rid of the toxic conditions. We cannot get rid of these acids because the kidneys, bowels, lungs, and skin are not throwing off the body acids fast enough. When these acids are not thrown off quickly enough, they stay in the body, becoming so active that acidemia and other irritated conditions arise. A person with a highly acid stomach and acid reactions in the body will find that he is allergic to many foods. Citric acid would not produce as many irritating effects in persons with this problem if they would first make sure that the eliminative organs were working properly.

Lemons, and all citric acid fruits, are good in cases of putrefaction, especially of the liver. In many cases, they will help stir up any latent toxic settlements in the body that cannot be eliminated any other way. Lemon drinks help tremendously when we need to remove the impurities and the fermentative effects of a bad liver. We have often used citric acid diets with excellent results. But citrus juices do thin the blood, and we must remember that the elimination diet is only part of what we require for right living.

Lemons are wonderful for throat trouble and catarrh. At the first sign of a cold, drink a glass of warm, unsweetened lemonade, and the cold may be prevented. Lemons may aid in digestion and can strengthen resistance. A little lemon and the yolk of a raw egg in a glass of orange juice is an excellent mild laxative, as well as a nutritious drink. But, if you are extremely irritable, nervous, sensitive, or highly toxic, use vegetable juices and vegetable broths instead of the citric acid fruits.

Lemons are wonderful for fevers, because a feverish body responds to citric acid fruits better than any other food. If we could live correctly, we would find that citrus fruits are one of the most wonderful foods to put into the body. By "living correctly," I mean that if the skin is eliminating properly, it would be able to take care of its share of the waste materials that have to be eliminated. When the skin is not eliminating well and acids are stirred up with citrus fruit, the kidneys have to do more work than they are capable of doing. In this case it is best to use vegetable juices instead of citrus juice to avoid stirring up the toxemia acids in the body. Vegetable juices carry off toxemia acids and act more as a sedative. Before we use lemons we should make sure that the eliminative organs are working well, because if they are not, the citric acid will cause overactivity. This overactivity will result in constant catarrhal discharges, as well as many highly acid reactions in the body.

Lemons can be used very effectively in cases of influenza. My late teacher, Dr. V. G. Rocine, gave me this remedy for influenza many years ago: Bake a lemon for twenty minutes in the oven. Cut it in half and squeeze one half of the baked lemon into a glass of hot water. Drink this every half hour, as long as the fever is present.

The lemon seems to have the properties of increasing elimination through the skin, and therefore helps reduce the fever. The lemon also has certain effects on the germ life found in influenza, since it is a wonderful germicide. In fact, there are at least twenty different germs that can be destroyed by the use of the lemon itself. To make this influenza remedy more complete, Dr. Rocine used a boneset tea along with it to control the calcium that is necessary whenever there is fever.

NUTRIENTS IN ONE POUND (including peel)

Calories	90	Iron	3.1 mg
Protein	3.3 g	Vitamin A	30 I.U.
Fat	.9 g	Thiamine	0.06 mg
Carbohydrates	48.1 g	Riboflavin	0.18 mg
Calcium	274 mg	Niacin	0.9 mg
Phosphorus	67 mg	Ascorbic acid	346 mg

LENTIL

Lentils have been cultivated for thousands of years, and evidence that they were used in the Bronze Age has been found. They do not grow wild. Lentils are legumes, and their protein content is second only to soybeans. They contain as much protein as many muscle meats.

Lentils make a hearty, filling soup. When preparing them, simmer for one-and-a-half to two hours.

THERAPEUTIC VALUE

Lentils neutralize muscle acids in the body, and are especially good for the heart. They help build the glands and blood, and may be used with a variety of vegetables and grains in soups to provide a rich supply of minerals for nearly every organ, gland, and tissue in the body.

NUTRIENTS IN ONE POUND

Calories	1,542	Iron	30.8 mg
Protein	112 g	Vitamin A	270 I.U.
Fat	5.0 g	Thiamine	1.69 mg
Carbohydrates	272.6 g	Riboflavin	.99 mg
Calcium	358 mg	Niacin	9.3 mg
Phosphorus	1,710 mg	Ascorbic acid	_____

LETTUCE

Lettuce is one of the oldest vegetables and probably originated in India or Central Asia. According to the writings of Herodotus, lettuce was served to the Persian kings as far back as the sixth century B.C. It was a popular Roman food at about the beginning of the Christian era, and in the first century A.D. a dozen distinctly different varieties were described by Roman writers of that era. There is also evidence that lettuce was grown in China in the fifth century A.D.

Columbus may have carried lettuce seeds to the New World, for it was being cultivated in the Bahamas in 1494. It was a common vegetable in Haiti as early as 1565, and Brazil was reported to have been cultivating it before 1650. The early colonists evidently introduced lettuce into the United States, and in 1806 sixteen varieties were reported growing in American gardens.

Both the English and Latin words for lettuce are based on the heavy, milky juice of the vegetable, which is a characteristic of the lettuce family. The primitive forms of lettuce had long stems, and large leaves grew at the ends of these stems. The close-packed lettuce heads were well developed in Europe by the sixteenth century, while the loose-head type was developed later.

Lettuce has become the most valuable truck crop, and 85 percent of the commercial crop is produced in the West—California, Arizona, Colorado, Washington, Oregon, and Idaho. The Northeast and South Atlantic states are also important lettuce-growing regions.

Lettuce is available all year, and the peak months are May, June, and July. Although the Crisphead and Butterhead types of lettuce are the most important from a commercial standpoint, the Cos or Romaine type is best from a health standpoint, as the sun is allowed to penetrate each leaf. The leaves also generally have less of the bitterness that is characteristic of some types of head lettuce. The "leaf" or "bunching" type of lettuce is distinguished by loose leaves that do not form a head. This type is best for home gardening, as it can be grown in areas where the temperature is too high for successful growing of the other types of lettuce. The Stem type lettuce has an enlarged stem and no head. The leaves are not as palatable as the other types of lettuce leaves except when young and tender. The stems are pulled and eaten raw or cooked.

Lettuce of good quality should be fresh, crisp, and tender, and if in head-lettuce form, the head should be fairly firm to hard. Lettuce with a well-developed seed stem has a bitter flavor.

THERAPEUTIC VALUE

Leaf lettuce is much richer in iron than head lettuce. I do not advocate using head lettuce in the diet, for it contains little nourishment. It contains significantly lower amounts of vitamins A and C than green Romaine lettuce. The darker green outside leaves contain a much higher proportion of the valuable food elements than the light-colored inner leaves. Head lettuce is very gas-forming, and really only offers bulk to the intestinal tract. It has an alkaline ash, however, and is not stimulating. Also, it is excellent for those who would like to lose weight. It also has many sleep-promoting elements and makes good lettuce juice, which will help promote sleep. It tends to slow down the digestive effect of the intestinal tract.

NUTRIENTS IN ONE POUND (head lettuce)

Calories	57	Iron	1.6 mg
Protein	3.8 g	Vitamin A	1,710 I.U.
Fat	0.6 g	Thiamine	0.20 mg
Carbohydrates	0.1 g	Riboflavin	0.21 mg
Calcium	86 mg	Niacin	0.5 mg
Phosphorus	78 mg	Ascorbic acid	24 mg

LIMA BEAN

Records found in old Peruvian tombs show that lima beans have been around for centuries. European explorers found this vegetable in Lima, Peru, and this is where the name comes from. Lima beans probably originated in Guatemala, and are still grown in tropical regions.

The flourishing dry lima bean industry of southern California seems to have started in 1865. In this year, Henry Lewis bought a few hundred pounds of lima bean seeds from a tramp steamer from Peru that had put in port at Santa Barbara. Most of the dry lima bean crop is produced along the Pacific coast from Santa Ana to Santa Barbara, and Florida is also a large producer of lima beans. The peak months of supply are July through October.

There are two types of lima beans. The large "potato" type have large pods and are fleshy and not likely to split at maturity. The baby lima bean is an annual plant that matures early. The pods are small and numerous, and are likely to split open at maturity.

When selecting lima beans, look for quality pods that are fresh, bright green in color, and well-filled. Lima beans, when shelled, should be plump with tender skins, green to greenish white. The skin should puncture when it is tested. Hard, tough skins mean that the bean is overmature, and these beans usually lack flavor. Lima beans are often called "butter" beans.

THERAPEUTIC VALUE

Lima beans can be used either dry or fresh. Fresh lima beans are alkaline and have a high protein value. Dry limas are hard to digest, and the dry skin is irritating to an inflamed digestive system. Lima beans are beneficial to the muscular system.

Lima beans are excellent as a purée in soft diets for stomach disorders. They make a tasty baked dish, such as bean loaf. One pound of lima beans contains as many nutrients as two pounds of meat!

Dry beans have a high protein content of almost 18 percent, but fresh beans are only 4 percent protein. The kidney bean and navy bean are very similar in makeup and therapeutic value to the lima bean.

NUTRIENTS IN ONE POUND (unshelled)

Calories	234	Iron	4.2 mg
Protein	13.6 g	Vitamin A	520 I.U.
Fat	1.5 g	Thiamine	0.38 mg
Carbohydrates	42.8 g	Riboflavin	0.21 mg
Calcium	115 mg	Niacin	2.5 mg
Phosphorus	288 mg	Ascorbic acid	48 mg

LIME

The lime is native to southeastern Asia and has been cultivated for thousands of years. It is believed that the Arabs brought limes with them from India during the period of Mohammedan expansion in A.D. 570–900. However, the lime was not mentioned by historians until the time of the Crusades. In 1626, Sir Thomas Herbert spoke of finding oranges, lemons, and limes on the island of Mohelia, off Mozambique. From the earliest days of British sailing vessels, British sailors were given a regular ration of lime juice to prevent scurvy at sea, resulting in the nickname "limey" for British sailors. Lime trees grew on the island of Haiti as early as 1514, and the cultivated lime spread from the West Indies to Florida. Later, it was even found growing spontaneously in thickets or as scattered plants. The so-called "wild lime groves" found on the lower East Coast, on the Florida Keys, were really planted by Henry Perrine in 1838. Congress had granted him land for the growth of economical tropical plants.

Limes have been grown in California and Florida since the early days of the citrus industry. After the great freeze in Florida in 1894–95, when the lemon industry was almost totally destroyed, California began growing virtually all the lemons in the United States. At this time Florida's lime industry expanded, and now Florida grows most of the limes used in this country. California is second in production, and Mexico is a close third.

Limes grow all year. Florida produces them from April to April, and California from October throughout the year. The main season for imports is May through August.

The principal variety of lime grown in this country is the large-fruited acid lime, which has few or no seeds. This variety, called the

Florida Persian, has a smooth, tight rind, and is a light orange-yellow color when ripe. Its pulp is fine-grained, tender, and a light greenish-yellow. It is very acid and highly flavored. California's Bearss lime is of this same type. The Mexican lime is lemon-yellow when ripe, with a smooth, tight rind; has greenish-yellow, fine-grained, tender pulp, with abundant, strong-flavored juice; and is very acid. There is a sweet-type lime, but it is not grown in this country.

Limes that are green in color and heavy for their size are the most desirable commercially, because of their extreme acidity. The full, ripe, yellow lime does not have a high acid content. If the lime is kept until fully ripe it may be used in the very same way the lemon is used, and to fortify other foods with vitamin C. Like lemons, limes are very high in vitamin C, are a good source of vitamin B_1, and are rich in potassium. They spoil easily, and limes with a dry, leathery skin or soft, moldy areas should be avoided. Store limes in a cool, dry place.

Limes contain 5 to 6 percent citric acid, and are too acid to drink without sweetening. Their natural flavor is enhanced when combined with other juices. Limes make a delicious dressing for fish, and, when added to melons, bring out the natural flavor of the melon. A few drops of lime juice added to consommé, or jellied soups, give a particular zest to the flavor. Subacid fruits, such as apples, pears, plums, peaches, grapes, and apricots, go best with limes.

THERAPEUTIC VALUE

Limes are good for the relief of arthritis because they have such a high vitamin C content. They are especially good for anyone with acidemia, because they are one of the most alkalinizing foods. A drink of lime juice and whey is a wonderful cooler for the brain and nervous system. Limes can be used to treat brain fever, or someone who is mentally ill. They are good for a brain with a great deal of hot blood in it, which usually shows itself in anger, hatred, or other brain disturbances. Limes make a wonderful sedative for those suffering from these afflictions.

NUTRIENTS IN ONE POUND (without rinds or seeds)

Calories	107	Iron	2.3 mg
Protein	2.8 g	Vitamin A	50 I.U.
Fat	0.8 g	Thiamine	0.10 mg
Carbohydrates	42.4 g	Riboflavin	0.08 mg
Calcium	126 mg	Niacin	0.7 mg
Phosphorus	69 mg	Ascorbic acid	94 mg

MANGO

The mango is said to have originated in Burma, Malaya, or the Himalayan region of India. It has been in cultivation for over 4,000 years and has entered prominently in Hindu mythology and religious observances. It is now a familiar fruit to all parts of the tropic zone, and is as important there as the apple is in our more temperate climate.

Although the mango is not too well-known in this country, some parts of the world value this fruit highly. Glowing descriptions of mangos can be found in the literature of these countries. The Turkoman poet, Amir Khusrau, for instance, wrote of the mango in the fourteenth century: "The mango is the pride of the garden, the choicest fruit of Hindustan. Other fruits we are content to eat when ripe, but the mango is good at all stages of growth."

The first attempt to introduce the mango into this country was made in 1833, when plants were transported to Florida from Mexico. These trees died, and another attempt was made thirty years later when seedling trees were introduced. The real success of its culture came at the beginning of this century, when choice grafted trees were brought from India. Because of the fruit's susceptibility to frost, its culture is limited to certain sections of Florida, where it is a summer crop only.

The mango tree is a member of the sumac family. It sometimes grows as high as forty feet. Its leaves are shiny and its flowers yellow or of a reddish hue. There are hundreds of varieties of mangos, and they range from the size of plums to that of apples, often weighing a pound or more. The common color of the mango is orange, although the fruit may range from green to yellow or red.

This fruit is available from May to September, the peak month being June. Some varieties are shipped in from China, Jamaica,

Mexico, and Cuba. A quality mango has a fairly small seedstone, and the pulp is delicate and smooth. The fruit should be fresh in appearance, plump, and firm to the touch; however, the test of quality is in its taste.

Mangos are best eaten as a fresh fruit. They have a high sugar content, although they are slightly acid in taste. Mangos are good used in combination with other fruits in salads, and in some parts of the world they are roasted. Both the flavor and aroma of mangos are spicy and attractive. To conserve the aroma, do not cut until just before serving.

THERAPEUTIC VALUE

Mangos contain a considerable amount of gallic acid, which may be binding to the bowels. It is excellent as a disinfectant to the body. Many people claim the mango is a great blood cleanser, and it also has fever-soothing qualities. Mango juice will reduce excessive body heat. Mangos are also wonderful for helping to throw off body odors.

NUTRIENTS IN ONE POUND

Calories	198	Iron	0.6 mg
Protein	2.1 g	Vitamin A	14,590 I.U.
Fat	0.6 g	Thiamine	0.19 mg
Carbohydrates	51.6 g	Riboflavin	0.17 mg
Calcium	27 mg	Niacin	2.8 mg
Phosphorus	39 mg	Ascorbic acid	106 mg

MELON

The many varieties of the popular melon give us certain elements not found in any other food. The honeydew melon originated in Asia, and it is believed that as early as 2,400 B.C. this distinct type of muskmelon was growing in Egypt. The cantaloupe is native to India and Guinea, and has been cultivated for more than 2,000 years. In Europe, it was first grown from seed transported from its native habitat.

The highly alkalinizing honeydew was introduced to America in 1900, and Arizona and California have become the biggest producers. It is available the year around, but it is at its peak of abundance in July through September. The cantaloupe is available from late May through September, but is most abundant in June and July.

Both the honeydew and the casaba, which is another variety of winter melon, are usually picked before maturity and ripened off the vine. Cantaloupe, however, do not develop any additional sugar after they are picked. This melon should be picked when it is still hard and pulls off the vine smoothly, without leaving a jagged scar.

Learn to select melons by the color and firmness of their rind, and by fragrance. The cantaloupe may have a coarse netting over its surface, or it may be of fine texture, depending again upon variety. Choose cantaloupe for their sweet fragrance. The casaba rind is golden in color and should feel heavy when ripe. A ripe honeydew has a creamy yellow surface color, and usually the scar in the blossom end yields to slight pressure.

The coloring of the flesh also is important, both as to degree of ripeness and to pleasing the eye and thus the palate. When fully ripe, casaba melons are cream in color, honeydews a yellowish-cream in color, and cantaloupes either a light or dark shade of salmon, depending upon variety. Deeply colored flesh in the melon denotes that it will be high in vitamin A.

It is important to pick a thoroughly ripe watermelon in order to receive the greatest benefit. A ripe watermelon, when thumped with the fingers, has a dull, hollow sound. Another test of a good ripe melon is to try to scrape the rind with the fingernail; when the green skin comes off easily, the melon is ready to be eaten. Good watermelon has firm, crisp, juicy flesh and is never dry or fibrous.

Melons are very high in silicon, especially if eaten right down to the rind. When we discard watermelon rind, we are missing one of its greatest elements. To obtain the gland- and blood-building chlorophyll, run the rind through a liquifier or juicer.

Watermelon, of course, is well-known as an efficient eliminator. Because it has such a high content of water and soluble chemicals, it can go into the bloodstream quickly and reach many of the organs of the body, depositing the chemicals needed to carry away waste.

During the melon season, we should fortify the body against

the winter months with a "melon reserve" of vitamins A, B and C, which are found in delightful form in the melon family

THERAPEUTIC VALUE

Melon gives us an excellent supply of distilled water that contains the finest mineral elements possible. Many of us think that we are drinking enough water, but our city water supplies do not give us "pure" water. Melons, with their root system, pick up water from deep, in-ground reserves, and bring it to our tables in a delicious fruit substance. Consider the melon for rejuvenation and alkalinizing the body. Melons also are excellent for aiding elimination.

NUTRIENTS IN ONE POUND

Calories	65	Iron	0.4 mg
Protein	1.0 g	Vitamin A	1,240 I.U.
Fat	0.4 g	Thiamine	0.10 mg
Carbohydrates	14.4 g	Riboflavin	0.11 mg
Calcium	15 mg	Niacin	0.4 mg
Phosphorus	25 mg	Ascorbic acid	13 mg

MULBERRY

The mulberry tree grows wild in many parts of the United States. The fruit can be black, white, red, or purple. The white variety is not as tasty as the other types, which are sweeter. Mulberries taste best in pies, jams, and desserts.

THERAPEUTIC VALUE

In times past, mulberries were believed to have great curative powers and were used as a general cure-all. Now, they are used mainly in eliminating and weight-loss diets. They are also good for building the blood, and for coating the stomach in any gastric disturbance. Mulberry juice is especially good for the digestive system. The mulberry is fairly high in vitamin B. It is very soothing to the nerves because of its high phosphorus content.

MUSHROOM

The Pharaohs of Egypt monopolized mushrooms for their own use. They thought they were too delicate to be eaten by common people. The Egyptian potentates did not understand the sudden, overnight appearance of mushrooms, and consequently believed they grew magically. By the first century B.C., the mushroom had gained such a fine reputation among epicures of the Roman Empire that the poet Horace celebrated its goodness in verse. The Romans called mushrooms "food of the gods," and served them on festive occasions. They were thought to provide warriors with unusual strength.

Up to the seventeenth century, only the wild types of mushrooms found growing in meadows and pastures were known. During the reign of Louis XIV, mushroom growing was introduced in France. Parisian market gardeners experimented to learn the secrets of successful mushroom culture. By 1749 mushroom beds were cultivated in caves and cellars, and the results were much better than when they were grown outdoors. The British were raising mushrooms in hothouses sometime before 1700.

The commercial production of mushrooms in the United States started in the late 1890s when a group of florists in Chester County, Pennsylvania started growing them under the benches in their greenhouses. The greatest event in the history of mushroom culture in the United States occurred in 1926 when a farmer found a clump of pure white mushrooms in a bed of uniformly cream-colored fungi. Most of the mushrooms grown today are descendants of this white clump.

Mushrooms are now cultivated in specially constructed buildings that are windowless and in which temperature and humidity are controlled. Mushroom spawn is cultivated by laboratory scientists who sell it to the growers for inoculation of the mushroom beds. Such precise methods are necessary to provide pure spawn of known characteristics.

The introduction of mushrooms into gravies, sauces, soups, and other dishes adds zest and flavor, but they also are a fine food when served as a vegetable. Mushrooms require very little preparation. Wash, cut off the bottom portion of the stem if it has dried, and either slice the caps and stems or leave whole, depending on the method of cooking. Butter a deep pan, cut up the mushrooms so they fill the pan to a depth of about two inches, and simmer over a low heat until the mushrooms are covered with their own juice.

This may take more than ten minutes. Then, cook more briskly for about five minutes, until tender. Overcooking toughens mushrooms.

Green plants can get their food by manufacturing it in their leaves from air, water, sunshine, and soil nutrients, but mushrooms cannot do this. They have no leaves, so they must depend on green plants to make their food for them, and they cannot use it unless it is in the process of decay. Mushrooms propagate from spores, a brownish powder shed from the rounded head which, when ripe, opens like a parasol. However, cultivated mushrooms are not reproduced from spores, but from fine strands of mycelium, which are rootlike growths that spread through organic material. Most wild mushrooms are not poisonous, but unless you know the difference, you should leave them alone. It is not possible to tell by taste which are dangerous. Some very unpalatable mushrooms are harmless, while others that have an agreeable taste are poisonous.

Scientists today say that darkness is not the primary requisite for growing mushrooms. They say that, for healthy growth, all mushrooms need constant temperature and protection against drafts.

The term mushroom refers to a large number of different species and varieties of fleshy fungi. Only one species is usually cultivated and that is *Agaricus Campestris*, which has a straight stem, a smooth cap of a shade varying from white or ivory to brown, and gills of different shades of pink. Most of the cultivated mushrooms grown in the United States are of the white variety variously known as Snow White, White King, White Queen, etc. This variety is very prolific and is preferred by nearly all markets because of its attractive, clean, white appearance.

THERAPEUTIC VALUE

Prior to the mid-1940s, all you needed to do to work up a hot argument among nutritionists was to say the word "mushrooms." Scientists' assertions about the food value of mushrooms ranged from calling them "vegetable beefsteak" full of proteins, to declaring that they had no protein and very little else. This confusion arose partly from the fact that mushrooms of many species were investigated and the results reported under a common head. A June 1946 report by William B. Eccelen, Jr. and Carl R. Fellers of the Massachusetts Agricultural Experiment Station, stated that cultivated mushrooms of the *Agaricus Campestris* type compare favorably in food value to many fresh fruits and vegetables.

Mushrooms are among the few rich organic sources of germanium, which increases oxygen efficiency of the body, counteracts the effects of pollutants, and increases resistance to disease. Because mushrooms are extremely low in calories, they are useful in reducing diets. They are also a good source of vitamin B.

NUTRIENTS IN ONE POUND

Calories	123	Iron	3.5 mg
Protein	11.9 g	Vitamin A	trace
Fat	1.2 g	Thiamine	0.41 mg
Carbohydrates	19.4 g	Riboflavin	2.02 mg
Calcium	26 mg	Niacin	18.6 mg
Phosphorus	510 mg	Ascorbic acid	14 mg

MUSTARD GREENS

The large-leafed pungent garden mustards grown in this country as pot herbs are generally the brown or Indian mustard. Many types of Indian mustard have been found over the middle half of Asia. There are several centers of development for this plant. The primary one is in Northwest India and adjacent areas, and secondary centers are in Eastern India, Assam, Burma, and China.

Young, tender leaves of mustard greens can be used as salad leaves, while the older, tender leaves are used as cooked greens. These should be cooked in a tightly covered pan for fifteen or twenty minutes with only the water that clings to the leaves after washing. Seasoning depends on taste.

Mustard plants are grown for their seeds in Montana, California, Washington, Oregon, and North Dakota. Indian mustards and other types are grown, some for medicinal purposes and some for use in condiments. In Russia, mustard seed oil has been used in place of olive oil.

Mustard greens can be mixed in salads or cooked gently with other vegetables. Mixing them with other vegetables helps cut down on their strong, biting taste.

THERAPEUTIC VALUE

Mustard greens compare favorably with other green, leafy vegetables in their nutritive content. They are superior to some in that they do not lose mineral values (such as calcium) through the presence of the nutrient-stealer, oxalic acid. Mustard greens also contain one of the B-complex vitamins (nicotinic acid), which is a preventive and cure for pellagra. Mustard greens are an alkaline food.

NUTRIENTS IN ONE POUND

Calories	98	Iron	9.6 mg
Protein	9.5 g	Vitamin A	22,220 I.U.
Fat	1.0 g	Thiamine	.30 mg
Carbohydrates	17.8 g	Riboflavin	.68 mg
Calcium	581 mg	Niacin	2.8 mg
Phosphorus	159 mg	Ascorbic acid	308 mg

NECTARINE

The history of the nectarine goes back to the early part of the Christian era, then merges with that of the peach. Sturtevant writes that the first mention of nectarines was made by Cieza de Leon in the mid-fourteenth century when he described the Caymito of Peru as "large as a nectarine." However, U. P. Hedrick is convinced that Pliny's "duracinus" (A.D. 79) is the nectarine. Since Dalechamp in 1587 and J. Bauhin in 1650 described nectarines, other botanists and pomologists have included them in their lists of fruits. In the sixteenth and seventeenth centuries the nectarine was called "nucipersica" because it resembled the walnut in smoothness and color of the outer skin as well as in size and shape. Parkinson in 1629 described six varieties and said, "I presume that the name Nucipersica doth most rightly belong unto that kind of Peach, which we call Nectorins . . ." Robert Beverly described them as most abundant in Virginia, in his *History of Virginia*, published in 1720. He further said that the Indians "had greater variety and finer sorts of them (peaches and nectarines) than the English." The word "nectarine" comes from the Greek "nekter," which is the drink of the gods in Greek and Roman mythology.

The nectarine is a smooth-skinned peach. Experiments show that nectarines may grow from peach stones, and peaches from nectarine stones. Peach trees can produce nectarines by bud-variation and nectarine trees also produce peaches, and the fruit so produced will come true to seed. Also, either peach or nectarine trees may produce a fruit half peach and half nectarine, and subsequently produce a true peach. The trees, leaves, and seeds of these fruits are indistinguishable. The characteristics of the fruits are the same except the nectarine has a fuzzless skin, is smaller, and has firmer flesh, greater aroma, and a distinct and richer flavor. Varieties in nectarines are parallel to those in peaches, being either clingstone or free, and the flesh may be red, yellow, or white.

Nectarines may be used in any of the ways peaches are used—fresh as a table fruit, stewed, baked, or made into preserves, jams, and ice cream. They can be canned and also dried.

In the humid eastern United States, nectarines are not as successfully grown as peaches. For this reason virtually the entire commercial crop is grown in California. Nectarines are on the market June through September from domestic sources and January through March from abroad.

The most important shipping varieties are: Quetta, which is a large, deep-colored clingstone fruit; John Rivers, a medium-sized, variety that is highly crimson on exposed cheek, and is practically a freestone; and Gower, a medium-sized, highly colored freestone fruit, which is the earliest commercial variety.

THERAPEUTIC VALUE

Nectarines are considered a subacid fruit and can be mixed with any fruit. They leave an alkaline ash, and are best eaten raw. They are wonderful dried.

NUTRIENTS IN ONE POUND

Calories	267	Iron	2.4 mg
Protein	2.1 g	Vitamin A	6,817 I.U.
Fat	trace	Thiamine	——
Carbohydrates	71.4 g	Riboflavin	——
Calcium	17 mg	Niacin	——
Phosphorus	100 mg	Ascorbic acid	54 mg

OKRA

Okra is native to tropical Africa, where it has been cultivated for many centuries. It is now widely grown in warm regions. For many years it has held an important place among the garden vegetables of the southern states.

The young and tender seed pods of okra are used to give a pleasant flavor and provide thickening for soups and stews. In Louisiana, okra is used in Creole cookery and is the "gumbo" used in many dishes. It is excellent also as a boiled vegetable. Just wash it, boil about ten minutes in salted water until tender, drain, and serve with butter or lemon butter. Okra and tomatoes make a fine combination. Raw sliced okra is good in salads. Okra should preferably be cooked in stainless steel, agate, porcelain, earthenware, or glass utensils. Copper, brass, iron, or tin will cause the okra to discolor, turn black, and look unappetizing.

Okra is a soft-stemmed annual of the mallow family and is closely related to the shrubby althea. It grows three to five feet high, and bears yellow flowers which are followed by fruiting capsules or seed pods.

There are three general types of okra: tall green, dwarf green, and ladyfinger. Each of these is again divided according to length and color of the pods. Varieties in most common use are known to the seed trade as Perkins Mammoth, Long Green, Dwarf Green, and White Velvet. Clemson Spineless is of the same type as Perkins Mammoth Podded but has spineless pods and somewhat sparse foliage, making it less troublesome to harvest than other varieties.

Young, tender, fresh, clean pods of small to medium size usually are of good quality. Pods should snap or puncture easily. Pods that have passed their prime look dull and dry. They are usually woody, and the seeds are hard. If held too long, they are likely to become shriveled and discolored, and lack flavor.

THERAPEUTIC VALUE

The sodium content of okra is very high. It also contains a vegetable mucin that is soothing to the irritated membranes of the intestinal tract. Okra has an alkaline reaction.

Okra is made into tablets, and they are valuable in replenishing a sodium deficiency in the body and in replacing sodium lost through excessive perspiration. The tablets are also good for ulcers of the stomach.

This low-calorie vegetable helps keep the joints limber. Okra powder is very good to include in broths and soups. Because it contains a high amount of sodium, it is good for elderly people.

NUTRIENTS IN ONE POUND

Calories	140	Iron	2.8 mg
Protein	9.4 g	Vitamin A	2,030 I.U.
Fat	0.8 g	Thiamine	0.49 mg
Carbohydrates	29.6 g	Riboflavin	0.42 mg
Calcium	328 mg	Niacin	2.8 mg
Phosphorus	199 mg	Ascorbic acid	121 mg

ONION

Onions are believed to have originated in Asia. When the Israelites were in the wilderness after being led out of Egypt by Moses, they yearned for onions and other vegetables they were used to eating. Onions were used by the Egyptians as offerings to their gods. They were fed to the workmen who built the pyramids, and Alexander the Great gave onions to his troops to promote their valor.

The odoriferous onion and the dainty lily are members of the same family, *Liliaceae*. The substance that gives the onion its distinctive odor and flavor is a volatile sulfurous oil which is about half eliminated by boiling. This volatile oil is what causes tears. Holding onions under cold water while peeling them prevents the oil fumes from rising, so use water and spare your handkerchief.

Onions lose approximately 27 percent of their original ascorbic acid (vitamin C) after five minutes of boiling.

There are two classes of onions—strong and mild. The early grown onions are generally milder in flavor and odor and are preferred for raw use. Each of these two classes can be again categorized into four colors—red, brown, white, and yellow. The white onions are the mildest. Each has many varieties.

Onions are also further divided by size for different uses. The smallest size is the pickling onion, also known as pearl or button onion, and is not more than one inch thick. The next size is the boiling onion, which is usually an inch to two inches in diameter.

The next larger size is preferred for chopping or grating. The very large Spanish or Bermuda onions are mild and sweet and good for slicing. They average two and one-half to two and three-quarters inches in diameter. In the trade, the term Valencia is used to mean Spanish-type yellow onions. The globe and flat-type yellow onions are generally referred to as yellows, and white onions of the globe and semi-globe types are generally referred to as whites.

Texas is the main early spring producer; California and Texas the main late spring states; California and New Jersey the most important early summer producers; and New York, Michigan, Minnesota, Colorado, California, Idaho, and Oregon the principal late summer states.

THERAPEUTIC VALUE

Onions are one of the earliest known food medicines, and were used for hundreds of years for colds and catarrhal disorders and to drive fermentations and impurities out of the system. The liquid from a raw onion that has been chopped up fine, covered with honey, and left standing for four or five hours, makes an excellent cough syrup. It is wonderful for soothing an inflamed throat. Onion packs on the chest have been used for years in bronchial inflammations.

Onions contain a large amount of sulfur and are especially good for the liver. As a sulfur food, they mix best with proteins, as they stimulate the action of the amino acids to the brain and nervous system. Whenever onions are eaten, it is a good idea to use greens with them. Parsley especially helps neutralize the effects of the onion sulfur in the intestinal tract.

NUTRIENTS IN ONE POUND

Calories	157	Iron	2.1 mg
Protein	6 g	Vitamin A	160 I.U.
Fat	0.4 g	Thiamine	0.15 mg
Carbohydrates	36 g	Riboflavin	0.10 mg
Calcium	111 mg	Niacin	0.6 mg
Phosphorus	149 mg	Ascorbic acid	38 mg

ORANGE

The citrus fruit is one of the oldest fruits known in the history of cultivation. As early as 500 B.C. the fruit of the citrus tree was mentioned in a collection of old documents believed to be edited by Confucius himself. In the year A.D. 1178, Han Yen-Chi, a Chinese horticulturist, wrote on the subject of oranges, and the seedless orange was mentioned in these writings. This author speaks of twenty-seven varieties of "very valuable and precious" oranges.

Oranges were originally brought from China to India, and gradually spread over the entire world where the climate was mild enough for their cultivation. The sour orange, or "Naranga," as it was referred to in Sanskrit about A.D. 100, came into cultivation in the basin of the Mediterranean long before the fall of the Roman Empire. The sweet variety, or "Airavata," does not appear to have been cultivated until early in the fifteenth century, and then became so popular that it was soon being cultivated extensively throughout Southern Europe. The Moors brought the Seville orange from the East.

Wild oranges were found in the West Indies and Brazil as early as 1600. The early Spanish explorers are believed to have brought oranges with them to this country in the time of Ponce de Leon's quest for the Fountain of Youth. In California, the orange was cultivated at the San Diego Mission in 1769 and, in the year 1804, 400 seedlings grew into a grove of considerable size around the San Gabriel Mission. The popularity of the orange, particularly in the favorable climate of California, grew rapidly, until it soon developed into a leading industry. The orange became known as "California's liquid sunshine."

The original orange was very small, bitter, and full of seeds, but through constant efforts in cross-fertilization and selection, many varieties of this delicious fruit are now cultivated with a tremendous improvement in the quality of the fruit. The sweet oranges are, by far, the most popular, while the sour orange is used more for its propagating stock than for its fruit. Unless killed by frost or fire, the orange tree lives to an old age and continues to bear fruit throughout its lifetime.

More than two hundred varieties of oranges are grown in the United States. In 1919 the United States produced only about 25 percent of the world's total output of oranges, but now it produces about half. Oranges comprise about 60 percent of the citrus fruit grown in the United States.

Oranges are available every day of the year, but are most abundant in the United States from January to May. California, Florida, and Texas are the orange-producing states, and each of these states ships great quantities. California's vast Valencia orange acreage is now more extensive than the Navel orange plantings. This state now has about 150,000 acres of Valencias, and about 100,000 acres of Navels, with an additional few thousand acres of miscellaneous orange varieties. The largest proportion of the California orange crop—about 85 to 90 percent—comes from southern California.

Choose the first oranges of the season, for they are the richest in mineral values. Tree-ripened oranges have, by far, the greatest mineral content. The best quality orange is firm and heavy, has a fine-textured skin varying in texture according to variety, and is well-colored. The light orange lacks juice. Avoid the soft, flabby, or shriveled orange and those oranges with any soft or moldy areas upon them. Do not eat unripe oranges because they can cause stomach upsets, particularly in small children. Once the skin is cut or broken, the fruit should be eaten immediately as the vitamin C is harmed by exposure to the air. If orange juice is kept for a period of time, store in the refrigerator in an airtight container.

The orange is classified as a subtropical fruit and has a citric acid content of 1.5 percent. This alkaline-reacting fruit is best eaten with other tropical or subtropical fruits, with acid fruits, or with nuts or milk. It is best to avoid eating this fruit with starches or sweets, or with dried fruits.

Use oranges as a dessert fruit, with yogurt, or in combination salads. Make a cup of a segmented orange—the thick-skinned seedless orange is best for segmenting—and fill with cottage cheese. Make liquefied drinks, mixing orange juice with other subtropical or tropical fruits such as cactus fruits, loquats, mango, papaya, persimmon, pineapple, pomegranate, apples, and citrus fruits. Many have advised eating oranges or drinking orange juice with meals, early in the morning on an empty stomach, or directly following a meal if the body is in a highly acid condition.

The orange is one of the best sources of water-soluble vitamin C. The absence or insufficiency of this causes scurvy. As vitamin C is the least stable of all the vitamins, storage of orange juice at low temperature destroys the vitamin to some extent, and sterilization may destroy it completely. Generally, I think it is best to use the citric acid fruits in sections rather than in juices. When the orange is eaten in sections, the mineral material found in the pulp will help to neutralize the citric acid effect as it goes into the body.

Citrus fruits are high in sodium, but only when completely matured in the sunshine. The fruit acids from green or immature fruit cause many adverse body reactions.

If the section and bulk of the orange is fresh and sweet, it is an excellent food for children as a supplement for those who must drink cow's milk, or any milk, because it seems to help in the retention of calcium in the body. Ripe oranges contain as much as 10 percent fruit sugar, which can be immediately assimilated by the body.

THERAPEUTIC VALUE

Oranges are the most popular source of vitamin C. They are excellent for treating overacid body conditions, constipation, or a particularly sluggish intestinal tract. In cases of acidosis, drink orange juice, or eat oranges after meals. If the intestinal tract is not functioning properly, drink a large glass of orange juice upon awakening in the morning, or about one-half hour before breakfast. In cases of stomach acid deficiency, start the meal with a peeled orange or a glass of orange juice.

Those who suffer from tooth decay or poor gums are probably lacking in vitamin C and should drink large amounts of orange juice for a period of a few weeks. People with gastric and duodenal ulcers are deficient in ascorbic acid, and their diet should be supplemented with a high potency vitamin C such as that found in fresh oranges and orange juice.

Oranges are very good for elimination. They stir up the acid accumulations and catarrhal settlements in the body very quickly. However, sometimes this is not a good idea if the channels of elimination, such as the skin and kidneys, are not able to take out these acids fast enough.

A body acid condition resulting from a high protein diet can best be neutralized and eliminated by the alkaline mineral elements found in fruits and vegetables. Oranges and all citrus fruits are particularly valuable because they are alkaline-reacting. People who are ill with a cold or other minor illness and who still must continue their daily work will find that orange juice, or a citrus juice diet, is the next best thing to a complete fast and rest in bed.

In cases of high fevers or arthritis, drink orange juice freely. Use orange juice where soft diets are required, and where a low calorie diet is necessary.

Eat the whole orange, excluding the very outer skin, to get all the good from the fruit. The luscious orange is rated tops in importance in the contribution to good health.

NUTRIENTS IN ONE POUND

Calories	164	Iron	1.3 mg
Protein	2.9 g	Vitamin A	910 I.U.
Fat	0.7 g	Thiamine	0.25 mg
Carbohydrates	36.6 g	Riboflavin	0.8 mg
Calcium	108 mg	Niacin	0.8 mg
Phosphorus	75 mg	Ascorbic acid	162 mg

PAPAYA

The papaya is native to Central America. From there it has been introduced to areas favorable to its growth in Asia, Africa, and Polynesia. It is second only to the banana in importance in South and Central America and Hawaii. The papaya tree is actually a large shrub, not unlike a palm in appearance, and bears fruit when it is only a few months old. The fruit resembles a melon with smooth skin, and is yellowish-orange in color when ripe. The flesh is a darker orange and is from one to two inches thick. In the center of the fruit are a large number of small, round, black seeds.

The papaya has been planted in Florida and Texas, where it has met with considerable success. In California its cultivation is confined to the most protected areas in the southern part of the state.

THERAPEUTIC VALUE

The papaya is rich in vitamins. It is especially high in vitamins A, C, and E, and is rich in calcium, phosphorus, and iron.

The papaya is high in digestive properties and has a direct tonic effect on the stomach. It is used in the treatment of stomach ulcers and fevers, and has a high mucus solvent action. The papaya retains its potency in high temperatures.

NUTRIENTS IN ONE POUND

Calories	119	Iron	.9 mg
Protein	1.8 g	Vitamin A	5,320 I.U.
Fat	.3 g	Thiamine	.12 mg
Carbohydrates	30.4 g	Riboflavin	.13 mg
Calcium	61 mg	Niacin	.9 mg
Phosphorus	49 mg	Ascorbic acid	170 mg

PARSLEY

It is believed that parsley had its beginning in the southern part of Europe, and has been grown in European gardens since the time of Charlemagne. History records that parsley was fed to the chariot horses in the days of the Roman Empire because it was believed that it would make them speedy. From very early times parsley was a supposed cure for many illnesses and was even sold for this purpose by travelers from Sardinia. This biennial plant had its beginning in this country during the earliest days of colonization.

There are two types of parsley: The foliage type, which is the most popular and is used for garnishing and flavoring, and the turnip-rooted type, which Europeans seem to prefer. This latter type is cooked and used like other root vegetables.

The South produces much of the parsley grown for commercial use, except in midsummer. Northern parsley for commercial use is grown from early spring until late autumn, with some production during the winter in the less cold regions, when coldframes are used.

Parsley may be obtained all year long. The peak months are June and July. A quality parsley is bright green, and free from dirt and yellowed or wilted leaves. It needs to be kept moist and cool to be fresh and inviting.

Most people use parsley as a flavoring or garnish. However, it is also good in vegetable juices and salads.

THERAPEUTIC VALUE

Parsley is a blood purifier and is good for stimulating the bowel. It has an alkaline ash. Parsley is high in iron and rich in copper and manganese.

When parsley is dried and used as a tea, it has a diuretic action. It is good for allaying kidney conditions, especially if no extreme inflammations exist, but too much of it could irritate the kidneys. Most kidney complaints will improve when parsley is added to the diet. Parsley is also good for the sexual system. It builds the blood and stimulates brain activity.

NUTRIENTS IN ONE POUND

Calories	200	Iron	28.1 mg
Protein	16.3 g	Vitamin A	3,298 I.U.
Fat	2.7 g	Thiamine	.54 mg
Carbohydrates	38.6 g	Riboflavin	1.19 mg
Calcium	921 mg	Niacin	5.6 mg
Phosphorus	286 mg	Ascorbic acid	780 mg

PARSNIP

Parsnips are believed to be native to the Mediterranean area and northeastward, including the Caucasus. The Romans believed that the parsnip had medicinal as well as food value. One story is that Emperor Tiberius imported them from Germany, where they grew in profusion along the Rhine. It is possible that the Celts of that part of Europe had brought the parsnip back from their forays to the east hundreds of years before. A German print, dated 1542, pictures the modern parsnip, and another shows it under the name of "pestnachen," a Germanized form of the old Roman "pastinaca."

By the mid-sixteenth century the parsnip was a common vegetable and was one of the staples of the poorer people of Europe, as the potato is today. Parsnips were well known by the first English colonists in America. They were grown in Virginia in 1609 and were common in Massachusetts twenty years later. Even the American Indians readily took up the growing of parsnips.

When properly cooked—and this means steamed, not boiled—parsnips have a sweet, nutty flavor. To obtain the full flavor of parsnips, they should be steamed in their skins until tender. Then they may be peeled and slit lengthwise. If the core is large, scoop it out with the point of a knife. The parsnips are then ready to be put through the ricer and served like mashed potatoes.

Smooth, firm, well-shaped parsnips of small to medium size are generally of best quality. Soft, flabby, or shriveled roots are usually pithy or fibrous. Softness is sometimes an indication of decay, which may appear as a gray mold or watery soft rot. Woody cores are likely to be found in large, coarse roots.

The parsnip is strictly a winter vegetable. Its flavor does not fully develop until it has been exposed to a temperature near freezing. Exposure to cold develops the sweet flavor. Scientists explain that at low temperatures the starch in parsnips gradually changes to sugar. At least two weeks exposure to a temperature around freezing is necessary for best flavor.

THERAPEUTIC VALUE

Parsnips are excellent for improving bowel action, and have a beneficial effect on the liver. They have a slight diuretic action and leave an alkaline ash in the body. Parsnips compare with carrots in food value. If tender, they can be eaten raw. They are considered a starch vegetable and a summer food.

NUTRIENTS IN ONE POUND

Calories	293	Iron	2.5 mg
Protein	1.8 g	Vitamin A	____
Fat	1.8 g	Thiamine	0.27 mg
Carbohydrates	64.4 g	Riboflavin	0.42 mg
Calcium	202 mg	Niacin	0.7 mg
Phosphorus	282 mg	Ascorbic acid	63 mg

PEA

Evidence shows that the pea has been around since prehistoric times. Although the pea is of uncertain origin, it is probably native to Central Europe or Central Asia. It is also probable that peas were brought from Greece or Italy by the Aryans 2,000 years before Christ.

Peas—pods and all—were considered a sovereign spring medicine in medieval England. At first, they were grown only for their dry seeds and, even today, some varieties are grown extensively for split pea soup. The green pea was not mentioned in historical writings until after the Norman conquest of England, and garden peas did not become common until the eighteenth century.

The green pea is a natural soluble mixture of starch and protein. Fresh peas are alkaline-forming, while dried peas have a tendency to produce allergic reactions and to cause gas, particularly when eaten with too much protein or concentrated starch.

The best quality pea is one that is young, fresh, tender, and sweet. Use fresh, young peas in order to obtain the greatest food value and flavor. The pod should be velvety soft to the touch, fresh in appearance, and bright green in color. The pods should be well filled and the peas well developed, but not bulging. The large ripe pea is really a seed and should not be considered a vegetable.

The real "sugar" pea is grown primarily in Europe and is little known in the United States. Because Chinese food is so popular in this country, there is a variety of pea grown and picked for the thick, soft, green pods that are used in these dishes. Their roughage is great for the intestinal tract, and they are very nourishing. However, this herbaceous, tendril-climbing legume can be eaten, pod and all, in any variety, if picked young enough. Those people who are troubled with a lot of gas or with a sensitive stomach wall or intestinal tract may find the hulls of the more mature pea irritating. In such cases, the peas should be puréed, or liquified, to avoid irritating disturbances.

Fresh green peas tend to lose their sugar content unless they are refrigerated to about 32°F shortly after being picked. They should be cooked soon after they have been picked, for they lose their tenderness and sweetness as they age. Shell just before cooking, retaining a few of the pods to cook with the peas for additional flavor. Cook in as little water as possible, so that no water need be discarded after cooking. If some pot liquor does remain after cooking, use in soup or as a base in the liquefied vegetable drink.

Never cook peas in bicarbonate of soda water in order to keep their fresh green appearance. This method not only destroys the food value and digestibility of the pea, but is totally unnecessary. Peas cooked in a vessel that is vapor-sealed or that has a tight lid, or steamed in parchment paper, with little water, retain their flavor, greenness, and vitamins. When combined with carrots or turnips, peas are particularly tasty, and when a little onion is added, they need not be seasoned. If seasoning is desired, add a little dehydrated broth powder after cooking and serve with butter.

There are over 1,000 cultivated varieties of peas listed in this country alone. However, only a few of these are commercially important. They may be classified generally as tall or dwarf, early or late, small pod or large pod. The most popular variety is probably the Alaskan, a tall, very early, small-pod pea. If planted in early May, this variety can be ready for table use in about sixty-two days. Many other varieties are almost equal in popularity, some of which are Little Marvel, Laxton, Gradus, and Giant Stride.

Green peas are available all year, but are particularly abundant March through July. California leads in production, with Florida, Texas, the Southern states, Idaho, Washington, Colorado, New York, and New Jersey also contributing an abundant supply of the nation's crops.

The pea is a fairly rich source of incomplete protein. As an alkaline ash vegetable, it is highly nutritious when eaten raw, and is more easily digested than beans. However, it takes a strong digestive tract to properly digest raw peas. To eat in their raw state, liquefy, and combine with other vegetables, proteins, or starches, to help aid in their digestion. Do not combine with fruits.

THERAPEUTIC VALUE

This alkaline-reacting vegetable is an outstanding source of vitamins A, B₁, and C. The pea pods are very high in chlorophyll, iron, and calcium-controlling properties. Discarded pods are discarded vitamins and valuable minerals.

Fresh garden peas are slightly diuretic in action. They also give relief to ulcer pains in the stomach because they help use up the stomach acids. In cases of ulcers, however, peas should be puréed.

Peas do not have any other particular healing properties. People who have a vitamin A deficiency should eat them raw, liquefied, or in juice. They should be eaten in combination with non-starchy vegetables to get the full value of the vitamin A they contain.

NUTRIENTS IN ONE POUND

Calories	201	Iron	3.9 mg
Protein	13.7 g	Vitamin A	1,390 I.U.
Fat	9.8 g	Thiamine	0.69 mg
Carbohydrates	36.1 g	Riboflavin	0.33 mg
Calcium	45 mg	Niacin	5.5 mg
Phosphorus	249 mg	Ascorbic acid	54 mg

PEACH

Peaches have been cultivated in both China and Persia since ancient times and are probably Chinese in origin. The peach has been found in Chinese writings as far back as the tenth century B.C. In the fifth century B.C., Confucius refers to the "tao," the Chinese word for peach, in his writings. This "Persian apple," as the peach was called, was introduced into Greece and Rome about A.D. 100. Then it was introduced to northern Europe, and soon it became one of the most popular of all fruits. In Europe, France is known as the principal peach producing country. The early settlers brought the peach to the United States, and it found the soil and climate so congenial that three centuries after it reached America some of our leading botanists believed it to be native to this country.

Peach growing on a commercial basis began on a small scale in the United States in the early nineteenth century. In 1870, a man by the name of S. R. Rumph, of Marshallville, Georgia, produced the Elberta from a seed of Chinese clingstone peaches. He gave a seed of the same lot to L. A. Rumph, who produced a variety that became known as the Belle of Georgia. The Elberta became so popular in the South that by 1889 it was placed on the approved list of varieties sponsored by the American Pomological Society. From 1910 to the present time the Elberta has held the lead in commercial production.

Forty states produce peaches commercially. Although Georgia is known as "the peach state," California leads in production, but most of its peach crop goes into cans. The peach-growing states are, in order of production: California, Georgia, Michigan, South Carolina, North Carolina, Arkansas, Washington, Colorado, Texas, Alabama, Tennessee, Illinois, Pennsylvania, New York, and New Jersey.

Peaches are on the market from late May to mid-October. The peak is in July and August. Georgia usually takes the lead in large-quality shipment.

A peach of fine quality is firm and free from blemishes. It has a fresh appearance, and is either whitish or yellowish in color, combined with a red color or blush, depending on the variety. If the peach is picked green or immature it will not ripen satisfactorily, and may develop a pale, weak color and will shrivel. The flesh will become tough and rubbery and will lack flavor. Peaches do not gain sugar after they are picked as they have no reserve of starch.

Peaches are wonderful in combination with other fruits in a

salad, and can also be mixed with vegetables. When eaten with other foods they are best with a protein meal. Cheese and peaches can be used in good combination, especially when traveling. They mix well with all dairy products.

THERAPEUTIC VALUE

Peaches, especially fruit that has a high color, are high in vitamin A. The peach has a high sugar and water content and is very laxative to the body. Peaches are wonderful in alkalinizing the blood stream, and they help stimulate the digestive juices. They can be used to regulate the bowel and build the blood.

Peaches are an excellent food for elderly people, because the body assimilates this food very easily. Because they are easy to digest, very ripe peaches can be eaten in cases of ulcers of the stomach or inflammation of the bowel, and in cases of colitis. For these conditions, peaches should be soaked, cooked, and puréed.

For those who are diet-conscious, peaches are wonderful in helping to eliminate toxins in the body, and they are good to eat on a weight-loss program. They make an ideal food with which to break a fast.

Peach leaves make an excellent tea that is wonderful as a cleanser for the kidneys. The stones of the peach have even been used in years past, broken up in broth, for their calcium content.

Most of the dried peaches sold today are processed in sulfur, and it is better to avoid them if good health is to be maintained. If you eat canned peaches packed in syrup, throw away the syrup and eat only the peach.

NUTRIENTS IN ONE POUND

Calories	150	Iron	2.4 mg
Protein	2 g	Vitamin A	5,250 I.U.
Fat	0.4 g	Thiamine	0.08 mg
Carbohydrates	38 g	Riboflavin	0.19 mg
Calcium	36 mg	Niacin	3.6 mg
Phosphorus	75 mg	Ascorbic acid	31 mg

PEAR

Pears were used as food long before agriculture was developed as an industry. They are native to the region from the Caspian Sea westward into Europe. Nearly 1,000 years before the Christian Era, Homer referred to pears as growing in the garden of Alcinous. A number of varieties were known prior to the Christian Era. Pliny listed more than forty varieties of pears. Many varieties were known in Italy, France, Germany, and England by the time America was discovered.

Both pear seeds and trees were brought to the United States by the early settlers. Like the apple, pear trees thrived and produced well from the very start. As early as 1771 the Prince Nursery on Long Island, New York, greatest of the colonial fruit nurseries, listed forty-two varieties. The introduction of pears to California is attributed to the Franciscan Fathers. Led by Father Junipera Serra, in 1776, they planted seeds carried from the Old World.

In the eighteenth and nineteenth centuries greatly improved pears were developed, particularly in Belgium and France. In 1850, pears were so popular in France that the fruit was celebrated in song and verse, and it was the fashion among the elite to see who could raise the best specimen. When the better varieties were brought into the United States a disease attacked the bark, roots, and other soft tissues of the trees, and practically destroyed the industry in the East. The European pear thrives primarily in California, Oregon, and Washington and in a few narrow strips on the south and east sides of Lake Michigan, Lake Erie, and Lake Ontario, where there are relatively cool summers and mild winters. Under these conditions, the trees are not as susceptible to pear blight, or "fire blight."

Another kind of pear, distinguished from the European "butter fruit" with its soft, melting flesh, had developed in Asia, and is known as the sand pear. These have hard flesh with numerous "sand" or grit cells. Sand pears reached the United States before 1840, by way of Europe, and proved resistant to fire blight. Hybrids of sand pears and European varieties are now grown extensively in the eastern and southern parts of the United States. They are inferior to the European pear, but still better to eat than the original sand pear. The best European varieties grow in the Pacific States, and from these states come most of the pears used for sale as fresh fruit for processing.

Pears are grown in all sections of the country, but the Western states (California, Oregon, and Washington), produce approximately 87 to 90 percent of all pears sold commercially. Practically all pears that are processed come from the Western states.

More than 3,000 varieties are known in the United States, but less than a dozen are commercially important today. The Bartlett outranks all other varieties in quantity of production and in value. It is the principal variety grown in California and Washington and is also the important commercial pear in New York and Michigan. It originated in England and was first distributed by a Mr. Williams, a nurseryman in Middlesex. In all other parts of the world it is known as Williams or Williams' Bon-Chretien. It was brought to the United States in 1798 or 1799 and planted at Roxbury, Massachusetts under the name of Williams' Bon Chretien. In 1817 Enoch Bartlett acquired the estate, and not knowing the true name of the pear, distributed it under his own name. The variety is large, and bell-shaped, and has smooth, clear yellow skin that is often blushed with red. It has white, finely grained flesh, and is juicy and delicious.

THERAPEUTIC VALUE

Pears have a fairly high content of vitamin C and iron. They are good in all elimination diets and are a wonderful digestive aid. They help normalize bowel activity.

Pears have an alkaline excess. They are a good energy producer in the winter, when used as a dried fruit, and are a delicious summer food when fresh.

NUTRIENTS IN ONE POUND

Calories	236	Iron	1.1 mg
Protein	2.6 g	Vitamin A	90 I.U.
Fat	1.5 g	Thiamine	0.8 mg
Carbohydrates	59.6 g	Riboflavin	0.16 mg
Calcium	49 mg	Niacin	0.5 mg
Phosphorus	60 mg	Ascorbic acid	15 mg

PEPPER, GREEN

The green pepper is very high in vitamin C. We get the benefit of this vitamin C if we eat fresh, raw peppers. They lose some of this vitamin C when cooked, but if properly steamed, not boiled or cooked over a high flame, we still get a great amount of good from peppers.

A green pepper adds zest and beauty to all sorts of green salads. A pepper cut in slices and filled with cream cheese makes a beautiful as well as nourishing food.

Green peppers are best eaten raw. They are good combined with apples, cheese, nuts, and dried fruits. Green peppers can also be stuffed with brown rice, meat substitutes, or meat itself. Raw green peppers in salads are good, too. If you find that the peppers are hard to get used to, try chopping them up in small pieces in your vegetable salads or in small strips and serve them with other fresh vegetables.

THERAPEUTIC VALUE

Peppers are classified as a protective food because they contain so many elements that build up resistance. They contain vitamin A, which makes our tissues more resistant, especially to colds and catarrhal infections in the respiratory organs, sinuses, ears, bladder, skin, and digestive tract. Vitamin A also promotes growth and the feeling of well-being. Vitamin B, also found in peppers, aids in food absorption and normalizes the brain and nervous system by increasing metabolic processes. Peppers are high in vitamin C, which is a wonderful health promoter as it wards off acidosis. The vitamin C in peppers compares to that of oranges and grapefruit.

The green pepper is high in silicon, and we need this element in our system to have beautiful hair, skin, nails, and teeth—it might well be called the "beauty element."

NUTRIENTS IN ONE POUND

Calories	95	Iron	1.5 mg
Protein	4.6 g	Vitamin A	2,410 I.U.
Fat	0.8 g	Thiamine	0.14 mg
Carbohydrates	21.7 g	Riboflavin	0.25 mg
Calcium	42 mg	Niacin	1.4 mg
Phosphorus	95 mg	Ascorbic acid	476 mg

PERSIMMON

For centuries Japan and China have been growing the Oriental or Japanese persimmon. It is probably native to China, since it was introduced to Japan from that country. The Japanese consider it their national fruit but it is more properly called Oriental rather than Japanese persimmon, since it is not native to Japan. Commodore Perry's expedition, which opened Japan to world commerce in 1852, is credited with the introduction of this fruit to the United States.

The persimmon that is native to the United States grows wild in the East from Connecticut to Florida, and in the West from Texas to Kansas. This persimmon is much smaller than the Oriental, but has richer flesh. The wild fruit grows in sufficient abundance to satisfy local demand, and little or no shipping is done.

In general, persimmons that have dark-colored flesh are always sweet and nonastringent and may be eaten before they become too soft. Varieties with light-colored flesh, with the exception of the Fuyu variety, are astringent until they soften. The astringency is due to the presence of a large amount of tannin, the same substance found in tea. As the fruit ripens and sweetens the tannin disappears. Ripening can take place just as well off the tree as on.

The Japanese remove the "pucker" from persimmons by placing them in casks that have been used for sake, or Japanese liquor. Allowing persimmons to sweeten naturally will remove the "pucker," or tannin.

The season for persimmons is October through December, and the peak month is November. Almost all commercial shipments originate in California. The Hachiya is the largest and handsomest oriental variety grown in this country. As a rule, California produces a seedless variety, but the Hachiya grown in Florida has one or more seeds. The Hachiya fruit is cone-shaped and terminates in a black point. The skin is a glossy, deep, orange-red and the flesh is deep yellow, astringent until soft, but sweet and rich when ripe. The Tanenashi is the more important variety in the southeastern states. There are many other varieties that are grown commercially.

Good quality fruit is well-shaped, plump, smooth, and highly colored. The skin is unbroken and the stem cap is attached. Ripeness is usually indicated by softness.

THERAPEUTIC VALUE

When thoroughly ripe, persimmons are a rich source of fruit sugar. Dried persimmons are almost as sweet as candy. They are rich in potassium, magnesium, and phosphorus, and are good to use in a soft diet.

NUTRIENTS IN ONE POUND

Calories	286	Iron	1.3 mg
Protein	2.6 g	Vitamin A	10,080 I.U.
Fat	1.8 g	Thiamine	.11 mg
Carbohydrates	73 g	Riboflavin	.08 mg
Calcium	26 mg	Niacin	.4 mg
Phosphorus	97 mg	Ascorbic acid	48 mg

PINEAPPLE

Pineapples were cultivated in the West Indies long before Columbus visited there. But after his voyage to the island of Guadaloupe, it was recorded in Spain that Columbus had "discovered" the fruit. The pineapple is native to tropical America and was known to the Indians as na-na, meaning fragrance, and to the Spanish explorers as piña, because of its resemblance to a pine cone.

History does not record how pineapples first reached Hawaii. For many years they grew wild. Then, a young Bostonian started commercial production of them there in 1901 on twelve acres of land. His company has enlarged to the present 25,000 acres.

The plant of this fruit grows from two to four feet high, with a rosette of stiff, sword-shaped leaves growing from its base. Out of the rosette center grows a single, fleshy, scaly-coated fruit that is four to ten inches long. A cluster of sword-shaped leaves surmounts the fruit.

Pineapples are grown in many parts of the world, but the United States is supplied principally from Cuba, Mexico, Hawaii, and Puerto Rico. They may be obtained all year long, but are most abundant from April through July. The peak months are June and July.

A ripe pineapple in quality condition has a fresh, clean appearance, a distinctive darkish orange-yellow color, and a decided fragrance. The "eyes" of the fruit are flat and almost hollow. If the fruit is mature it is usually heavier in proportion to its size. To test for ripeness, pull at the spikes. If they pull out easily, the fruit is ripe; discolored areas, or soft spots, are an indication of bruised fruit.

THERAPEUTIC VALUE

High in vitamin C, the pineapple is considered to be a protective fruit. It is wonderful for constipation and poor digestion. The pineapple helps digest proteins, and can be used in elimination diets. It leaves an alkaline ash in the body. Pineapple is thought to have a certain amount of iodine because it grows near the ocean.

When buying canned pineapple, make sure it is unsweetened. Pineapple goes well with fruit and nuts, and is good to eat on a fruit diet.

NUTRIENTS IN ONE POUND

Calories	123	Iron	1.2 mg
Protein	1 g	Vitamin A	170 I.U.
Fat	0.5 g	Thiamine	0.20 mg
Carbohydrates	33 g	Riboflavin	0.06 mg
Calcium	39 mg	Niacin	0.5 mg
Phosphorus	19 mg	Ascorbic acid	40 mg

PLUM AND FRESH PRUNE

The early colonists found plums growing wild along the entire Eastern coast. They were one of many fruits eaten by the Indians before the coming of the white man, and reports of early explorers mention the finding of plums growing in abundance. Today, however, native plums are not important commercially. The European type of plums, *Prunus Domestica*, has replaced the native plum. Plum pits from Europe probably were brought to America by the first colonists, for it is reported that plums were planted by the Pilgrims in Massachusetts, and that the French brought them to Canada.

Although plums came to America by way of Europe, they are believed to have originated in Western Asia in the region south of the Caucasus Mountains to the Caspian Sea. According to the earliest writings in which the European plum is mentioned, the species dates back at least 2,000 years.

Another species, *Prunus Institia*, known to us as the Damson plum, also came to America by way of Europe. This plum was named for Damascus and apparently antedates the European type, although Damson pits have been found in the lake dwellings of Switzerland and in other ancient ruins.

Another important species, the Japanese plum, was domesticated in Japan, but originated in China. It was introduced in the United States about 1870. This type is grown extensively in California.

Plums are grown in some of the Spanish mission gardens of California at least as early as 1792, and the first prune plums grown in California were produced at Santa Clara Mission. However, the present California prune industry is not based on these but on the French prune, *Petite Prune d'Agen*, scions of which were brought to California from France in 1856 by Pierre Pellier. French-type prunes grown in California orchards were shipped in to San Francisco markets in 1859.

Botanically, plums and prunes of the European or Domestica type belong to the same species. The interchangeable use of the term "plum" and "prune" dates back for several centuries. Plum is Anglo-Saxon, and prune is French. Originally they were probably synonymous. It is uncertain just when the word prune was first used to designate a dried plum or a plum suitable for drying. The prune is a variety of plum that can be dried without fermenting when the pit is left in. Fresh prunes, as compared with plums, have firmer flesh, higher sugar content, and, frequently, higher acid content. A ripe, fresh prune can be separated from the pit like a freestone peach, but a plum cannot be opened this way.

Of all the stone fruits, plums have the largest number and greatest diversity of kinds and species. H.F. Tysser, editor of *Fruit*

Manual, published in London, says there are over 2,000 varieties. Samuel Fraser, in his book *American Fruits*, speaks of a list of about 1,500 varieties of Old World plums alone, and says there probably are just as many varieties of plums native to this continent. In addition, there is a long list of Japanese and Chinese plums.

Almost all of the fresh plums that are shipped in the United States are grown in California. There are two types of California plums, Japanese and European. The former are marketed early in the season and the latter in midseason or later. The Japanese varieties are characterized by their large size, heart-shape, and bright red or yellow color. Japanese varieties are never blue.

Plums and prunes of good quality are plump, clean, of fresh appearance, full colored for the particular variety, and soft enough to yield to slight pressure. Unless one is well acquainted with varieties, color alone cannot be relied upon an an indication of ripeness. Some varieties are fully ripe when the color is yellowish-green, others when the color is red, and others when purplish-blue or black. Softening at the tip is a good indication of maturity. Immature fruit is hard. It may be shriveled and is generally of poor color and flavor. Overmature fruit is generally soft, easily bruised, and is often leaky.

THERAPEUTIC VALUE

Fresh plums are more acid to the body than fresh prunes. When too many plums are eaten, an overacid condition results. When prunes are dried, however, they are wonderful for the nerves because they contain a phosphorus content of nearly 5 percent.

Prunes have a laxative effect. The dried prune is better to eat than the fresh prune or plum. The salts contained in the dried prune are valuable as food for the blood, brain, and nerves. The French prunes are considered the best for their value to the nervous system.

NUTRIENTS IN ONE POUND

Calories	218	Iron	2.2 mg
Protein	3 g	Vitamin A	1,200 I.U.
Fat	0.9 g	Thiamine	0.28 mg
Carbohydrates	55.6 g	Riboflavin	0.18 mg
Calcium	73 mg	Niacin	2.1 mg
Phosphorus	86 mg	Ascorbic acid	20 mg

POMEGRANATE

Mohammed once told his followers: "Eat the pomegranate, for it purges the system of envy and hatred." The pomegranate is one of the oldest fruits known to man. Frequent references to it are found in the Bible and in ancient Sanskrit writings. Homer mentions it in his *Odyssey*, and it appears in the story of *The Arabian Nights*. The pomegranate is native to Persia and its neighboring countries, and for centuries has been extensively cultivated around the Mediterranean, spreading through Asia. King Solomon was known to have an orchard of pomegranates, and history speaks of the children of Israel wandering in the wilderness and remembering with longing the cooling taste of the pomegranate. Ancient Assyrian and Egyptian sculpture has depicted this fruit, and it is sometimes on ancient Carthaginian and Phoenician medals.

The word pomegranate is derived from the Latin word meaning "apple with many seeds." The fruit grows on a bush or small tree from twelve to twenty feet high. It grows to about the size of an orange or larger.

A pomegranate of good quality may be medium or large in size and the coloring can range from pink to bright red. The rind is thin and tough, and there should be an abundance of bright red or crimson flesh, with a small amount of pulp. The seeds are contained in a reddish, juicy pulp that is subacid and of fine flavor. They should be tender, easy to eat, and small in proportion to the juicy matter that surrounds them, while the juice should be abundant and rich in flavor.

There are many varieties of pomegranate. At least ten varieties were growing in southern Spain in the thirteenth century, as described by a writer of the time. It is a warm-climate fruit, and the leading producers in this country are California and the Gulf states.

This fruit will not mature in cooler climates, although there are dwarf forms grown in cool climates which have striking scarlet flowers that are sold commercially. Pomegranates are in season September through December, and October is the peak month.

THERAPEUTIC VALUE

Use only the juice of the pomegranate. This juice is one of the best for bladder disorders and has a slight purgative effect. For elderly people, it is a wonderful kidney and bladder tonic.

NUTRIENTS IN ONE POUND (edible portion)

Calories	160	Iron	.8 mg
Protein	1.3 g	Vitamin A	trace
Fat	.8 g	Thiamine	0.07 mg
Carbohydrates	41.7 g	Riboflavin	0.07 mg
Calcium	20 mg	Niacin	.7 mg
Phosphorus	.8 mg	Ascorbic acid	10 mg

POTATO

The potato is one vegetable that is abundant throughout the year. It comes in many varieties. Though called "Irish," the white potato is native to the mountains of tropical America from Chile to Mexico, and was widely cultivated in South America at the time of the Spanish Conquest. The Spaniards introduced the potato into Europe early in the sixteenth century, and it was Sir Walter Raleigh who showed England how to eat the potato with beef gravy. He, too, started the potato fad in colonial Virginia, but it was Sir Francis Drake who was supposed to have brought the potato to Ireland. The potato soon became second only to Indian corn as the most important food contribution of the Americas, and is now one of the most valuable vegetable crops in the world.

The potato is classed as a protective vegetable because of its high vitamin C content. It has been noted in the past that, as the potato became common, scurvy, which is prevalent where vitamin C is absent, became uncommon, and soon disappeared almost entirely in potato-eating countries.

I believe that if we had to confine ourselves to one food, the potato is the one on which we could live almost indefinitely, exclusive of other foods, as it is a complete food in itself. It was Professor Hinhede of Denmark, a food scientist during the last war, who proved to the world that a person could live on potatoes for a long period of time without any depreciation of body energy. In fact, he and his assistant lived three years solely on potatoes—raw and cooked. He not only proved the potato to be a complete food, but he also showed how inexpensive a diet it was at a cost of approximately only six cents a day. It is good, however, to eat potatoes with other vegetables; eating them by themselves may eventually cause constipation.

When selecting potatoes make sure they are smooth, shallow-eyed, and reasonably unblemished. Avoid the extra large potato as it may have a hollow or pithy center. Potatoes with a slight green color are sunburned and may have developed a bitter taste.

The energy value of the potato is approximately the same as bread, but it is a far better balanced food than bread, particularly in its content of potassium, iron, and vitamins C, B_1, and G. The potato is also lower in calories. Because potatoes are a starchy food, they put less work on the kidneys.

It is best to eat potatoes in as raw a form as possible. However, raw, cut potatoes should be eaten as soon as they are cut, as their oxidation is very rapid. I know of no other food that will turn green, ferment, and break down quicker than potatoes will when they have been juiced.

Potatoes may be sliced raw and used in salads. Juice them, mixed with parsley, beets, or other vegetables for flavor. Potato juice is a great rejuvenator and is a quick way to get an abundance of vitamin C as well as other vitamins and minerals. Why not munch on a raw potato? It is no more peculiar for a child to eat a piece of raw potato than it is for him to eat a raw apple.

Instead of throwing away the potato peeling, eat it, because it is rich in mineral elements. At least 60 percent of the potassium contained in the potato lies so close to the skin that it cannot be saved if the potato is peeled. Furthermore, potassium is a salt, and you do not need to salt potatoes if the potato peelings are used. If you feel you need more seasoning, use a mineral broth powder (dehydrated vegetables) instead of table salt. Even using sweet butter in place of salted butter is better, and is not difficult to get used to when the flavor is enhanced with the addition of broth powder.

There are numerous ways to prepare and serve potatoes. They have a bland flavor, so they can be used frequently in meals. It is best to cook potatoes on a low heat, if possible, and if they are not baked they should be cooked in a vapor-sealed vessel to retain their goodness. The art of cooking can be used to build or to destroy.

It is necessary that we realize the difference between a properly steamed potato and a boiled potato—one is alkaline and the other is acid. According to the Bureau of Home Economics, United States Department of Agriculture, when ordinary cooking methods are used, from 32 to 76 percent of the essential food values, minerals, and vitamins are lost due to oxidation, or are destroyed by heat or dissolved in water. In a vapor-sealed utensil, oxidation is practically eliminated, less heat is required, and waterless cooking is possible. The vitamins and minerals are preserved for you and are not carried away by escaping steam.

The outside of the potato is the positive side. The negative side is the inside. The inside is carbohydrate and is acid in body reaction. So, it is best, when making alkalinizing broths for example, that you discard the center of the potato before adding the potato to the broth ingredients. Throw this part of the potato into your garden if you have one and it will do its part to rebuild the soil.

In preparing potatoes for cooking, scrub and wash them thoroughly. Use a stiff brush to remove the dirt. To bake, drop them first in very hot water to heat them, then rub them with oil to keep their skins from getting too hard in the process of baking and to help them be more easily digested. Remember to bake them at a slow oven heat. In the last five minutes of baking raise the oven heat to about 400°F to break down the starch grains.

Before serving baked potatoes, they may be cut in half, scooped out, and mashed with nut butter, avocado, or a little grated cheese. Garnish with parsley or chives. Or, take plain, baked potatoes, split open, and serve with a Roquefort, cream, and chive dressing.

THERAPEUTIC VALUE

Potatoes leave an alkaline ash in the body, are low in roughage, and may be used in the treatment of acidosis. They can also be used for catarrhal conditions.

When trying to overcome catarrhal conditions, cut the potato peeling about a half-inch thick and use it in broth or soup, cooking very little. The resulting broth will contain many important mineral elements.

Potato soup can also be used to great advantage in cases of uric acid, kidney, and stomach disorders, and for replacing minerals in the system. To make potato soup, peel six potatoes, making sure the peelings are about three-quarters of an inch thick. Place in water in a covered kettle and simmer twenty minutes. Add celery to change the flavor if desired. Add okra powder if the stomach is irritated.

The potassium in the potato is strongly alkaline, which makes for good liver activation, elastic tissues, and supple muscles. It also produces body grace and a good disposition. Potassium is the "healer" of the body and is very necessary in rejuvenation. It is a good heart element also, and potatoes can be used very well in all cases of heart troubles.

Anyone with ailments on the left side of the body—the negative side, or the heart and intestinal side of the body—can use carbohydrates that are negative in character. Potatoes are one of the best negative foods to use for building up the left side of the body.

To use an old remedy, take slices of potatoes and use as a pack over any congested part of the body. This type of pack draws out static, toxic material, or venous congestion in any part of the body. Use a narrow, thumb-shaped piece of potato to help correct hemorrhoid conditions.

To control diarrhea, cook potato soup with milk. The milk controls the diarrhea—it has a constipating effect, if boiled. The potato adds bulk, which is also necessary to control this trouble.

I believe that raw potato juice is one of the most volatile juices and the strongest juice that can be taken into the body. It is used in many cases of intestinal disorders, as well as for rejuvenation.

NUTRIENTS IN ONE POUND (raw and pared)

Calories	279	Iron	2.7 mg
Protein	7.6 g	Vitamin A	trace
Fat	0.4 g	Thiamine	0.40 mg
Carbohydrates	6.8 g	Riboflavin	0.15 mg
Calcium	26 mg	Niacin	4.4 mg
Phosphorus	195 mg	Ascorbic acid	64 mg

PUMPKIN

The pumpkin, along with other squashes, is native to the Americas. The stem, seeds, and parts of the fruit of the pumpkin have been found in the ruins of the ancient cliff dwellings in the southwestern part of the United States. Other discoveries in these ruins indicate that the pumpkin may even have been grown by the ''basket makers,'' whose civilization precedes that of the cliff dwellers, and who were probably the first agriculturists in North America.

Present varieties of pumpkin have been traced back to the days of Indian tribes. One variety, the Cushaw, was being grown by the Indians in 1586.

Botanically, a pumpkin is a squash. The popular term pumpkin has become a symbol, or tradition, at Halloween and Thanksgiving. This tradition dates as far back as the first colonial settlers.

Pumpkin can be served as a boiled or baked vegetable and as a filling for pies or in custards. It also makes a good ingredient for cornbread.

Pumpkins are grown throughout the United States and are used mostly in or near the producing area. They are classed as stock feed and pie types, some serving both purposes. The principal producers are Indiana, Illinois, New Jersey, Maryland, Iowa, and California. They may be found in stores from late August to March, the peak months being October through December.

Pumpkins of quality should be heavy for their size and free of blemishes, with a hard rind. Watch for decay if the flesh has been bruised or otherwise injured. Decay may appear as a water-soaked area, sometimes covered with a dark, mold-like growth.

THERAPEUTIC VALUE

Pumpkins are very high in potassium and sodium and have a moderately low carbohydrate content. They are alkaline in reaction and are a fair source of vitamins B and C. Pumpkins are good in soft diets.

Pumpkin can be used in pudding or it can be liquefied. One of the best ways to serve pumpkin is to bake it. Pumpkin seeds and onions mixed together with a little soy milk make a great remedy for parasitic worms in the digestive tract. To make this remedy, liquefy three tablespoons of pumpkin seeds that have been soaked three hours, one-half of a small onion, one-half cup soy milk, and one teaspoon honey. Take this amount three times daily, three days in a row.

NUTRIENTS IN ONE POUND (without rind and seeds)

Calories	83	Iron	2.5 mg
Protein	3.8 g	Vitamin A	5,080 I.U.
Fat	0.3 g	Thiamine	0.15 mg
Carbohydrates	20.6 g	Riboflavin	0.35 mg
Calcium	66 mg	Niacin	1.8 mg
Phosphorus	138 mg	Ascorbic acid	30 mg

RADISH

The radish is a member of the mustard family, but is also related to cabbage, cauliflower, kale, and turnips. After this vegetable was introduced into Middle Asia from China in prehistoric times, many forms of the plant were developed. Even before the Egyptian pyramids were built, ancient records in Egypt indicate that radishes were a common food in that country. An ancient Greek physician wrote an entire book on radishes. In 1548, records show that people in England were eating radishes raw with bread or as a meat sauce. It is believed that Columbus introduced radishes to the Americas. They were seen in Mexico about 1500 and in Haiti in 1565, and were among the first vegetables grown by the colonists in this country.

Radishes are a cool season crop and, with long days and high temperatures, they go to seed. They can grow in all types of soil but may require readily available plant food. Some of the many varieties mature in twenty-one days, but the winter variety takes about sixty days. Radishes are marketed all year around, and the peak period is April through July. In the Orient, varieties are grown only for cooking, and in Egypt and the Near East a variety is grown for the green tops only. The American varieties, however, can be used for both roots and tops in salads, and cooked. Some of the larger varieties that are grown for pickling, cooking, and drying in the Orient are now being experimented with on the West Coast by Oriental gardeners, but they are used chiefly as a local product.

A good-quality radish is well-formed, smooth, firm, tender, and crisp, with a mild flavor. The condition of the leaves does not always indicate quality, for they may be fresh, bright, and green, while the radishes may be spongy and strong, or the leaves may be wilted and damaged in handling, while the radishes themselves may be fresh and not at all pithy. Old, slow-growing radishes are usually strong in flavor, with a woody flesh. Slight finger pressure will disclose sponginess or pithiness.

THERAPEUTIC VALUE

Radishes are strongly diuretic and stimulate the appetite and digestion. The juice of raw radishes is helpful in catarrhal conditions. The mustard oil content of the radish makes it good for expelling gallstones from the bladder.

A good cocktail can be made with radishes. This cocktail will eliminate catarrhal congestion in the body, especially in the sinuses. It will also aid in cleansing the gall bladder and liver. To make this cocktail, combine one-third cucumber juice, one-third radish juice, and one-third green pepper juice. If desired, apple juice may be added to make this more palatable.

An excellent cocktail for nervous disorders is made from radish juice, prune juice, and rice polishings. This drink is high in vitamin B and aids in the flow of bile.

NUTRIENTS IN ONE POUND

Calories	49	Iron	2.9 mg
Protein	2.9 g	Vitamin A	30 I.U.
Fat	.3 g	Thiamine	.09 mg
Carbohydrates	10.3 g	Riboflavin	.09 mg
Calcium	86 mg	Niacin	.9 mg
Phosphorus	89 mg	Ascorbic acid	74 mg

RASPBERRY

The red raspberry was first cultivated about 400 years ago on European soil. Cultivation spread to England and the United States, where the native American raspberry was already well known.

In 1845, Dr. Brinkle of Philadelphia became the first successful producer of raspberries in this country, and he originated many varieties. By 1870, this berry had become an important crop in the United States.

The red raspberry is native to the northern United States, and the black raspberry is found in the South. The purple raspberry is a hybrid between the red and the black, and did not become important until about 1900.

The raspberry has a wide range of colors. A yellow raspberry is found growing wild in many areas, particularly in Maryland. The Asiatic species of raspberry has a color that ranges through red, orange, yellow, lavender, purple, wine, to black. Even white berries are found in many species in their wild state. Pink berries have been found in Alabama and Oregon, and lavender ones in North Carolina. In the West, the wild black raspberry is often not quite black, but rather a deep wine in color. The market berry is usually the cultivated berry and is both red and black. There are many varieties of each that are popular. The market supply runs from mid-April through July, and the peak month is July.

A quality berry is plump, with a clean, fresh appearance, a solid, full color, and is usually without adhering caps. Berries with caps attached may be immature. Overripe berries are usually dull in color, soft, and sometimes leaky.

THERAPEUTIC VALUE

Raspberries are considered a good cleanser for mucus, for catarrhal conditions, and for toxins in the body. They are a good source of vitamins A and C. Raspberries leave an alkaline reaction. They should never be eaten with sugar.

Raspberries are wonderful in juice form and can be used as a cocktail before meals, since they stimulate the appetite. Raspberry juice is delicious mixed with other juices.

NUTRIENTS IN ONE POUND

Calories	294	Iron	4.1 mg
Protein	5.4 g	Vitamin A	trace
Fat	6.2 g	Thiamine	0.12 mg
Carbohydrates	59.9 g	Riboflavin	30 mg
Calcium	132 mg	Niacin	1.3 mg
Phosphorus	97 mg	Ascorbic acid	81 mg

RUTABAGA

The rutabaga, believed to be a hybrid of the turnip and some form of cabbage, has a much more recent origin than the turnip. Some authorities say rutabagas are native to Russia and Siberia, others that the new species was first found in Europe some time in the late Middle Ages and probably originated in Scandinavia. It is said there was no record of the rutabaga until 1620 when the Swiss botanist, Casper Bauhin, described it. The word "rutabaga" comes from the Swedish "rotabagge." "Rota" means "root." It is often called Swedish turnip.

The rutabaga was known in the United States in about 1800 as "turnip-rooted cabbage." Both white and yellow-fleshed rutabagas have been known in Europe for more than three hundred years. In England, rutabagas were grown in the royal gardens in 1664. Botanically, the rutabaga is a member of the same genus that cabbage, kale, cauliflower, Brussels sprouts, collards, broccoli, turnips, and others belong to, the genus *Brassica*. However, it is a distinct species.

The roots of the rutabaga are either globular or elongated, and the leaf has no hair and is fleshier than the turnip. Also, the rutabaga has a longer ripening period. The turnip has flat roots, hairy leaves that are not fleshy, and the plants take less time to mature. Additionally, the rutabaga has much denser flesh than the turnip and is somewhat higher in total dry matter and total digestible nutrients. Although there are white-fleshed and yellow-fleshed varieties of both turnips and rutabagas, most rutabagas are yellow-fleshed and most turnips are white-fleshed. A further important difference is that the rutabaga grows best where the weather is colder, and is principally cultivated in the northern latitudes.

Rutabagas should be firm and fairly smooth, with few leaf scars around the crown and with very few fibrous roots at the base. Soft or shriveled rutabagas are undesirable, because they will be tough when cooked. Avoid roots that are light for their size, as they are likely to be tough, woody, pitted, or hollow and strong in flavor.

THERAPEUTIC VALUE

Rutabagas are sometimes recommended for cases of constipation. However, because of their mustard oil content, they are apt to cause gas. They should not be used by anyone who suffers with kidney troubles. Rutabagas contain more vitamin A than turnips.

Rutabagas should be steamed or stewed. They are a starch food, and this should be considered when combining them with other foods.

NUTRIENTS IN ONE POUND

Calories	177	Iron	1.5 mg
Protein	4.2 g	Vitamin A	2,240 I.U.
Fat	0.4 g	Thiamine	0.29 mg
Carbohydrates	42.4 g	Riboflavin	0.30 mg
Calcium	254 mg	Niacin	3.6 mg
Phosphorus	150 mg	Ascorbic acid	166 mg

SALSIFY

Salsify is a perennial herb that grows wild in Europe and Asia. It is also called common viper's grass or the oyster plant. Wild chicory and bitter chicory are often improperly categorized as salsify.

The salsify root may be prepared the same way as parsnips. It is quick-cooking and can be served with different sauces. The tender leaves can be used in salad.

THERAPEUTIC VALUE

Salsify contains a natural insulin that, when digested by the stomach's hydrochloric acid, helps take place of the starch digestion that the pancreas would have to do. It has an alkaline ash.

NUTRIENTS IN ONE POUND (with tops)

Calories	119	Iron	3.2 mg
Protein	6.2 g	Vitamin A	20 I.U.
Fat	1.3 g	Thiamine	.07 mg
Carbohydrates	38.4 g	Riboflavin	.09 mg
Calcium	100 mg	Niacin	.6 mg
Phosphorus	141 mg	Ascorbic acid	23 mg

SNAP BEAN

Both green and yellow snap beans are said to be native to South America and were introduced into Europe in the sixteenth century. There are over 150 varieties of these beans in cultivation today. Florida is the main producer of these beans, and California and other southern states also produce large quantities. Snap beans are available throughout the year, but their peak months are November through January, and April through June.

Good quality snap beans will snap readily when broken. The most desirable are pods with immature seeds. They should be well-formed, bright, fresh, young, and tender. Beans that have a dull or wilted appearance and whose seeds are half-grown or larger should be avoided, as they will probably be tough or woody.

Beans should not be overcooked, because this destroys the food value as well as the fresh green color. They can be run through a juice machine to make a very healthy juice. This juice may be more palatable when combined with other vegetable juice.

THERAPEUTIC VALUE

Beans, especially green beans, are an alkaline vegetable. The yellow or wax bean is inferior to the green bean in its chemical value to the blood, but is good to eat occasionally. Beans are high in protein, and are essential on a vegetarian diet.

NUTRIENTS IN ONE POUND (including inedible ends and strings)

Calories	128	Iron	3.2 mg
Protein	7.6 g	Vitamin A	2,400 I.U.
Fat	.8 g	Thiamine	.33 mg
Carbohydrates	28.3 g	Riboflavin	.42 mg
Calcium	224 mg	Niacin	2.0 mg
Phosphorus	176 mg	Ascorbic acid	76 mg

SOYBEAN

The soybean has grown extensively in Asia for hundreds of years. Now, the United States is the largest producer of soybeans, supplying two-thirds of the world's soybeans. The soybean is becoming more and more important, as people become aware of its nutritive value.

There are hundreds of varieties of soybeans. The soybean is a food that is rich in protein and minerals, and is the cheapest source of vegetable protein. Soy milk is similar to cow's milk but is much lower in fat; soy flour is good for making flat breads and breadsticks; and soy meal is valuable as livestock feed. However, the main product extracted from soybeans is soybean oil.

Soybeans are often used as a meat substitute, but I believe that this use has been overrated. They do have a high protein content, but are also high in starch; a person using them as an exclusive protein may become starch-logged. Soy milk, however, will not cause this reaction.

THERAPEUTIC VALUE

The soybean is valued for its high protein content, but it also has an exceptionally high amount of many other vitamins and minerals. Soybean oil will become more and more popular for both industrial and home use, since it is not only excellent for frying, is easily digestible, and contain no cholesterol, but also rejuvenates glands. Soy has an alkaline reaction.

NUTRIENTS IN ONE POUND (mature bean, dry, raw)

Calories	1,828	Iron	38.1 mg
Protein	154.7 g	Vitamin A	360 I.U.
Fat	80.3 g	Thiamine	4.99 mg
Carbohydrates	152.0 g	Riboflavin	1.43 mg
Calcium	1,025 mg	Niacin	10.1 mg
Phosphorus	2,513 mg	Ascorbic acid	_____

SPINACH

Spinach is native to Iran and adjacent areas, but did not spread to other parts of the world until the beginning of the Christian Era. The first record, written in Chinese, states that spinach was introduced into China from Nepal in A.D. 647. It reached Spain about A.D. 1100, brought by the Moors from North Africa, where it had been introduced by way of ancient Syria and Arabia. The prickly-seeded form of the vegetable was known in Germany in the thirteenth century and was commonly grown in European monastery gardens by the fourteenth century. A 1390 cookbook for the court of Richard II had recipes for "spynoches." The smooth-seeded form was described in 1552. Spinach was probably brought to the United States early in colonial days, but commercial cultivation did not start until about 1806 and the first curly-leaved variety was introduced in 1828.

Spinach (*Spinacia Oleracea*) is a small, fleshy-leaved annual of the goosefoot family. It is a quick-maturing, cool season crop that is hardy and will live outdoors over winter throughout most of the area from New Jersey southward along the Atlantic Coast and in most parts of the lower South. There are two other plants called spinach, but they are not genuine: New Zealand spinach and Mountain spinach, or garden orach. The former, sometimes called ice plant, is a small annual of the carpetweed family. It is chiefly an Australasian and Japanese herb used as a substitute for spinach. Mountain spinach also belongs to the goosefoot family. In the western United States it is part of the vegetation referred to as greasewood and is sometimes call saltbush.

Spinach has been both praised and abused. It has been popularized in the comic strips by the herculean feats of Popeye the sailor. On the other hand, Dr. Thurman B. Rice of the Indiana State Board of Health says, "If God had intended for us to eat spinach he would have flavored it with something." But flavoring is a job for cooks. The way spinach is thrown in a pot with a large quantity of water and boiled for a half hour or more, it's a wonder even Popeye relished it. Spinach should be cooked in a steamer with very little or no added water other than that clinging to the leaves after washing. If you insist on boiling it, again use only the water clinging to the leaves after washing, and cook in a covered pan for not more than ten minutes.

THERAPEUTIC VALUE

Spinach is an excellent source of vitamins C and A, and iron, and contains about 40 percent potassium. It leaves an alkaline ash in the body. Spinach is good for the lymphatic, urinary, and digestive systems.

Spinach has a laxative effect and is wonderful in weight-loss diets. It has a high calcium content, but also contains oxalic acid. This acid combines with calcium to form a compound that the body cannot absorb. For this reason, the calcium in spinach is considered unavailable as a nutrient. This is of small importance, however, in the ordinary diet. The oxalic acid factor would become important only if a person relied largely on spinach for calcium. The only effect the acid would have is if a large quantity of spinach juice were taken. This might cause disturbing results in the joints.

NUTRIENTS IN ONE POUND

Calories	89	Iron	13.6 mg
Protein	10.4 g	Vitamin A	26,450 I.U.
Fat	1.4 g	Thiamine	0.50 mg
Carbohydrates	14.5 g	Riboflavin	0.93 mg
Calcium	368 mg	Niacin	2.7 mg
Phosphorus	167 mg	Ascorbic acid	167 mg

SQUASH

Squash is native to the Western Hemisphere and was known to the Indians centuries before the arrival of the white man. It is a member of the cucurbit family, which includes pumpkins and gourds as well as cucumbers, muskmelons, and watermelons. Squash as we know it today is vastly different from the kind the Narragansett Indians dubbed "askutasquash," meaning "Green-raw-unripe"— which, incidentally, was the way they ate it. We still follow their example and eat summer squash while tender and unripe, though it is usually cooked.

Squash is best when steamed or baked; some people even use it in soup. The Hubbard squash, due to its hard shell, is usually baked in the shell. Squash may be used to add variety to the menu. Summer squash is boiled or steamed and served as a vegetable with drawn butter or cream sauce, or it may be served mashed. The delicate flavor of summer squash is lost by boiling it in large quantities of water and, of course, nutrients are lost when the cooking water is thrown away.

Squash may be grouped in five general types: Hubbard, Banana, Turban, Mammoth, and Summer. The latter are actually pumpkins. However, they are listed as squashes because that is what they are called in the market.

Summer squash should be fresh, fairly heavy for its size, and free from blemish. The rind should be so tender that it can be punctured very easily. Hard-rind summer squash is undesirable because the flesh is likely to be stringy and the seeds and rind have to be discarded. Winter squash should have a hard rind. Soft-rind winter squash is usually immature, and the flesh may be thin and watery when cooked, and lack flavor.

THERAPEUTIC VALUE

Winter squash contains more vitamin A than summer squash. Both are low in carbohydrates and can be used in all diets. Squash is a high potassium and sodium food that leaves an alkaline ash in the body. It is very good for the eliminative system.

NUTRIENTS IN ONE POUND (summer squash)

Calories	83	Iron	1.8 mg
Protein	4.8 g	Vitamin A	1,800 I.U.
Fat	1.0 g	Thiamine	0.23 mg
Carbohydrates	18.5 g	Riboflavin	0.38 mg
Calcium	123 mg	Niacin	4.5 mg
Phosphorus	128 mg	Ascorbic acid	75 mg

NUTRIENTS IN ONE POUND (winter squash)

Calories	161	Iron	2.0 mg
Protein	5.0 g	Vitamin A	11,920 I.U.
Fat	1.0 g	Thiamine	0.16 mg
Carbohydrates	39.9 g	Riboflavin	0.35 mg
Calcium	71 mg	Niacin	1.9 mg
Phosphorus	122 mg	Ascorbic acid	43 mg

STRAWBERRY

The strawberry is native to North and South America. An early Chilean variety was taken to Peru in 1557 and this same variety is still growing in Chile, Peru, Ecuador, and other South American countries. The modern strawberry was developed in Europe.

Most strawberry varieties that grow commercially today have originated within the last fifty-five years. Territories for their growth have expanded to almost every state in the Union, including the interior of Alaska.

How the name "strawberry" first came into use is often disputed. One researcher tells us that it was because straw was used between the rows to keep the berries clean and to protect the berries in the winter. Another explanation is that in Europe ripe berries were threaded on straws to be carried to market.

In 1945, about fifteen varieties constituted 94 percent of the total commercial market. The leading variety in the United States is the Blakemore, which originated in Maryland in 1923. Its firmness, earliness, and the fact that it holds its color when stored make it a leading market berry. The Klondike is grown extensively in Southern California and is one of the best shipping varieties. The Klonmore is native to Louisiana. Because it appears earlier, it is more resistant to disease and is fast replacing the Klondike in that state. Other popular varieties are the Howard 17 and the Marshall, which both originated in Massachusetts.

Strawberries are at their peak of abundance in April, May, and June; January, February, March, and July are moderate months.

Quality strawberries are fresh, clean, and bright in appearance. They have a solid red color, and the caps are attached. Strawberries without caps may have been roughly handled or are overmature.

THERAPEUTIC VALUE

Strawberries are a good source of vitamin C, and contain a large amount of fruit sugar. They are an excellent spring tonic, and are delicious when juiced.

They can be considered an eliminative food, and are good for the intestinal tract. Strawberries have an alkaline reaction in the body. Because of their high sodium content, they can be considered ''a food of youth.'' They also have a good amount of potassium.

Many people complain about getting hives from strawberries. This is usually because they are not ripened on the vine. If you are allergic to strawberries, try this: run hot water over them, then immediately follow this by running cold water over them. This takes the fuzz off the outside of the berries, which is believed to be the cause of the hives.

The seeds of the strawberry can be irritating in cases of inflammation of the bowel or colitis.

NUTRIENTS IN ONE POUND

Calories	179	Iron	3.5 mg
Protein	3.5 g	Vitamin A	250 I.U.
Fat	2.6 g	Thiamine	0.13 mg
Carbohydrates	35.3 g	Riboflavin	0.29 mg
Calcium	122 mg	Niacin	1.3 mg
Phosphorus	118 mg	Ascorbic acid	261 mg

SWEET POTATO

The sweet potato should be thought of as a true root and not a tuber, as is commonly believed. It has been one of the most popular foods of tropical and subtropical countries for centuries. Columbus and his men were fed boiled roots by the natives of the West Indies, which these men described as "not unlike chestnuts in flavor." This new food was carried back to Spain, and from there it was introduced to European countries. De Soto found sweet potatoes growing in the gardens of the Indians who lived in the territory that is now called Louisiana.

During the Civil War, troops short of rations found they could live indefinitely on sweet potatoes alone. The Japanese on Okinawa could not have held out as long as they did if they had not been able to raid sweet potato patches at night. In 1913 the supply of sweet potatoes was so large and the demand so small that Louisiana towns sold them for fifty cents a barrel.

There are two main types of sweet potatoes; those that are mealy when cooked, and those that are wet when cooked—popularly miscalled "yams." Actually, there are few yams grown in this country, and they are grown almost solely in Florida.

Decay in sweet potatoes spreads rapidly and may give the entire potato a disagreeable flavor. This decay may appear in the form of dark, circular spots or as soft, wet rot, or dry, shriveled, discolored and sunken areas, usually at the ends of the root.

Use the sweet potato baked, steamed, or roasted, in puddings or pies. Whenever possible, they should be cooked in their jackets, to conserve the nutrients. If you wish to discard the skin, this vegetable is much easier to peel when cooked. When combining the sweet potato with other foods, remember that it is a little more difficult to digest than the white potato.

THERAPEUTIC VALUE

The sweet potato is good for the eliminative system, but is a little more difficult to digest than the white potato. It contains a great deal of vitamin A and is a good source of niacin.

NUTRIENTS IN ONE POUND

Calories	419	Iron	2.7 mg
Protein	6.2 g	Vitamin A	30,030 I.U.
Fat	1.5 g	Thiamine	0.37 mg
Carbohydrates	96.6 g	Riboflavin	0.23 mg
Calcium	117 mg	Niacin	2.8 mg
Phosphorus	173 mg	Ascorbic acid	77 mg

SWISS CHARD

Swiss chard is a member of the beet family. Unlike most members of this family, chard does not develop an enlarged, fleshy root. Instead it has large leaves with thickened midribs, and both ribs and leaves are edible. The roots are hard and woody.

Swiss chard is a temperate zone biennial that withstands rather severe winters. It is of the same species as garden beets, mangel-wurzels, and sugar beets, and readily intercrosses with them through airborne pollination.

Chard is the beet of the ancients. Aristotle wrote about red chard and Theophrastus mentioned light-green and dark-green types of chard in the fourth century B.C. The Romans called this plant "beta," and the Arabs called it "selg." But chard was used as a potherb in the Mediterranean lands, Asia Minor, the Caucasus, and the Near East, long before Roman times. Wild beets grow widely in these areas.

Beets of the type that produce large, fleshy, edible roots were unknown before the Christian era. The ancients apparently used the root of the wild beet or chard for medicinal purposes only. Chard has been used in Europe for as long as there are definite records of food plants.

THERAPEUTIC VALUE

Swiss chard contains a great deal of vitamin C, vitamin A, potassium, sodium, and calcium. It is best not to cook it for a long time, because its vitamin content will decrease.

This vegetable is low in calories and high in alkaline ash. It is good when combined with other vegetables in salads, and helps ward off colds. It is beneficial to the digestive system, because it contains many of the vitamins and minerals essential to its operation.

NUTRIENTS IN ONE POUND (including 14 percent inedible parts)

Calories	82	Iron	9.8 mg
Protein	5.5 g	Vitamin A	10,920 I.U.
Fat	0.8 g	Thiamine	0.22 mg
Carbohydrates	17.2 g	Riboflavin	0.28 mg
Calcium	410* mg	Niacin	1.7 mg
Phosphorus	140 mg	Ascorbic acid	148 mg

*Calcium may not be available for body use because of the presence of oxalic acid.

TANGERINE

The tangerine is a citrus fruit that has been in cultivation in Southeastern Asia for probably more than 4,000 years. The name "tangerine" is supposed to be derived from Tangier, Morocco, and includes all the red-skinned mandarin oranges. They are all looseskinned, peeling as easily as a glove off your hand—thus the name "kid-glove" oranges was started in Florida in 1870 by Colonel George L. Dancy.

Sir Abraham Hume introduced the mandarin orange into England in 1805, and by 1850 it was well known in Italy. Between the years of 1840 and 1850, the Italian Consul brought the mandarin orange to Louisiana, planting it in the consulate grounds at New Orleans. From there the orange was taken to Florida, and in 1871, Colonel Dancy of Buena Vista, Florida began cultivation of the Dancy tangerine, the only tangerine now extensively growing there. The Clementine, which originated in the gardens of the orphanage of Misserghin, in Algeria, and was named in honor of Brother Clement, is also gaining in popularity in the United States, as is the Ponken, a famous Oriental fruit.

The season for marketing the tangerine is from November through May, and the peak months are December and January. A considerable amount of this fruit is on the market in November, February, and March.

Quality tangerines should be heavy for their size, which indicates ample juice content. They should have the characteristic deep orange or almost red color.

Mature fruit usually feels puffy because of the looseness of the skin.

THERAPEUTIC VALUE

Tangerines are high in vitamin C. The thin membrane that covers the segments contains a digestion-aiding factor, and should always be eaten. They have the same therapeutic value as oranges.

NUTRIENTS IN ONE POUND

Calories	160	Iron	1.3 mg
Protein	2.6 g	Vitamin A	1,360 I.U.
Fat	1 g	Thiamine	0.22 mg
Carbohydrates	35.1 g	Riboflavin	0.08 mg
Calcium	106 mg	Niacin	0.8 mg
Phosphorus	74 mg	Ascorbic acid	99 mg

TOMATO

It is believed that the present type of tomato is descended from a species no larger than marbles, that grew thousands of years ago. The tomato is native to the Andean region of South America and was under cultivation in Peru in the sixteenth century at the time of the Spanish conquest. Before the end of the sixteenth century, the people of England and the Netherlands were eating and enjoying tomatoes. The English called it the "love apple," and English romancers presented it as a token of affection; Sir Walter Raleigh is said to have presented one to Queen Elizabeth.

M. F. Corne is credited with being the first man to eat a tomato. His fellow citizens of Newport, Rhode Island, erected a monument to him, because the tomato was considered poisonous until Mr. Corne dared to eat one.

By cultivation and use the tomato is a vegetable; botanically, it is a fruit, and can be classified as a berry, being pulpy and containing one or more seeds that are not stones. It is considered a citric acid fruit and is in the same classification as oranges and grapefruit. Some oxalic acid is also contained in the tomato.

Consumption of tomatoes is on the increase. They are the third most important vegetable crop on the basis of market value; the first is potatoes. Tomatoes are produced in all states. In order of importance, the producers are: Texas, California, Florida, Ohio, and Tennessee. In the first four months of the year heavy shipments are imported from Mexico and Cuba. Fresh tomatoes are available all year, either from domestic production or imports. June and August are the peak months.

Tomatoes number greatly in variety, but it is estimated that only sixteen varieties are included in 90 percent of all tomatoes grown in the United States. Their characteristic colors range from pink to scarlet. A white tomato has recently been developed that is supposed to be acid-free. A good, mature tomato is neither over-ripe nor soft, but well developed, smooth, and free from decay, cracks, or bruises. Spoiled tomatoes should be separated immediately from the sound ones or decay will quickly spread.

If fresh, ripe tomatoes are unavailable, canned tomato and canned tomato juice are fine substitutes. It is preferable to use tomato purée, rather than canned tomatoes put up in water. Purée contains more vitamins and minerals.

Tomatoes are best when combined with proteins. Use tomatoes in both fruit and vegetable salads. They are cooling and refreshing in beverages, and are especially good as a flavoring for soups. Tomatoes can be used to give color, and make green salads more inviting.

Tomato juice should be used very soon after it has been drawn from the tomato, or after the canned juice is opened. If it is opened and left that way, it will lose much of its mineral value, because it oxidizes very quickly.

Tomatoes should be picked ripe, as the acids of the green tomato are very detrimental to the body and very hard on the kidneys. Many of the tomatoes today are grown in hothouses and are picked too green and allowed to ripen on their way to the markets or in cold storage plants built for this purpose. If the seeds, or the internal part of the tomato, is still green, while the outside is red, this is an indication that the fruit has been picked too green.

THERAPEUTIC VALUE

The tomato is not acid forming; it contains a great deal of citric acid but is alkaline forming when it enters the bloodstream. It increases the alkalinity of the blood and helps remove toxins, especially uric acid, from the system. As a liver cleanser, tomatoes are wonderful, especially when used with the green vegetable juices.

In many of the sanitariums in Europe tomatoes are used as a poultice for various conditions in the body. There is a mistaken belief that tomatoes are not good for those who have rheumatism and gout. People with these conditions should mix tomato juice with other vegetable juices to avoid a reaction that may be too strong.

Whenever the blood is found to be stagnant in any part of the body, a tomato poultice is wonderful as a treatment in removing that stagnation. It acts as a dissolving agent or solvent.

Tomatoes are very high in vitamin value. They are wonderful as a blood cleanser, and excellent in elimination diets. However, they should not be used to excess on a regular basis. Tomato juice can be used in convalescent diets, in combination with other raw vegetable juices such as celery, parsley, beet, and carrot juice.

NUTRIENTS IN ONE POUND

Calories	97	Iron	2.7 mg
Protein	4.5 g	Vitamin A	4,080 I.U.
Fat	0.9 g	Thiamine	0.23 mg
Carbohydrates	17.7 g	Riboflavin	0.15 mg
Calcium	50 mg	Niacin	3.2 mg
Phosphorus	123 mg	Ascorbic acid	102 mg

TURNIP AND TURNIP GREENS

The turnip, which belongs to the mustard family, is reported to have come from Russia, Siberia, and the Scandinavian peninsula. It has been used since ancient times. Columella wrote in A.D. 42 that two varieties of turnips were grown in what is now known as France. Pliny refers to five varieties, and stated that the broadbottom flat turnip and the globular turnip were the most popular.

Back in the sixteenth century, giant turnips created comment. In 1558, Matthiolus spoke of having heard of long purple turnips weighing thirty pounds; however, this may be considered small compared with the turnip weighing one hundred pounds grown in California in 1850.

Cartier sowed turnip seed in Canada as early as 1540, and they were cultivated in Virginia in 1609, and in Massachusetts as early as 1629. In 1707 they were plentiful around Philadelphia, and their use was recorded in South Carolina as early as 1779.

Turnips may be served steamed, with drawn butter or cream sauce. They are also excellent raw and shredded in salads.

Turnip greens are excellent cooked the same way spinach is usually cooked. The greens should be cooked in a covered pan until tender, using only the water that clings to the leaves.

Regardless of variety, turnips have much the same flavor if grown under the same conditions. They may be distinguished by shape, as round, flat, or top-shaped, and also by color of the flesh—white or yellow—by the color of the skin, and by the leaves. Varieties like Seven Top and Shogoin are grown almost exclusively for the leaves.

The most popular variety is the Purple Top White Globe. This variety has a large globe-shaped root, with an irregularly marked purple cap, and its flesh is white, sweet, crisp, and tender. The leaves are dark green, large, and erect.

THERAPEUTIC VALUE

Turnips are very high in sulfur and are sometimes gas forming. The root vegetable can be considered a carbohydrate vegetable. If eaten raw, they have a high content of vitamin C. Turnip juice is especially good for any mucous and catarrhal conditions. They have been used successfully in all bronchial disturbances, even asthma. Turnip packs over the chest are good for relieving bronchial disorders and packs over the throat are good for sore throats. When fresh and young, turnips can be used raw in salads. They leave an alkaline ash, and have a low calorie content and low carbohydrate content. They can be used in most diets.

Turnip leaves are considered good for controlling calcium in the body, as are all other greens. They have been used successfully in the South to combat pellagra, which is a disease caused by lack of calcium in the body.

NUTRIENTS IN ONE POUND (root vegetable)

Calories	117	Iron	2 mg
Protein	3.9 g	Vitamin A	trace
Fat	0.8 g	Thiamine	0.16 mg
Carbohydrates	25.7 g	Riboflavin	.26 mg
Calcium	152 mg	Niacin	2.2 mg
Phosphorus	117 mg	Ascorbic acid	140 mg

NUTRIENTS IN ONE POUND (turnip greens only)

Calories	140	Iron	9.1 mg
Protein	11 g	Vitamin A	34,470 I.U.
Fat	1.5 g	Thiamine	0.37 mg
Carbohydrates	20.6 g	Riboflavin	2.15 mg
Calcium	987 mg	Niacin	2.9 mg
Phosphorus	190 mg	Ascorbic acid	519 mg

WATERCRESS

Watercress is a member of the mustard family, which includes cabbage, kale, and broccoli. It is believed these foods originated in the Eastern Mediterranean and Asia Minor areas.

Watercress, common in Europe, North America, and lower South America, is an aquatic perennial that grows in regions that have small natural streams and limestone. The plants thrive when submerged in fresh running water, and there is no danger of winter-killing as long as the water does not freeze solid. Watercress grows in moist soil, usually along the banks of streams, and in recent years has been grown in greenhouses. Partial shade, moist soil, high humidity, and lime result in satisfactory growth. It is grown for its small, round, pungent leaves, which are eaten raw as salads or as garnishes, and as an ingredient in soups. Because of its flavor, watercress makes a tangy seasoning agent.

Brazilians have used watercress for treating tuberculosis, and experiments have reportedly shown improvement in a number of patients. Supposedly, the bacteria in watercress destroys the tuberculosis bacteria. Natives of Brazil have long produced a syrup which is claimed, in some cases, to be a remedy for this disease. This syrup is made by placing alternate layers of moist watercress and sugar in an earthen jar and burying it for fifteen to twenty days. After the liquid settles, the resulting syrup is supposed to be palliative and curative.

THERAPEUTIC VALUE

Watercress is a very alkaline food, and is most effective on a reducing diet. It is one of the best foods for taking care of catarrhal conditions and for purifying the blood. Watercress makes an excellent addition to vegetable juices.

Watercress is high in alkaline salts and the vitamins essential to warding off catarrhal conditions. It is good for glandular secretions and for the liver. Watercress is high in water content, so it is a wonderful dissolver. It is very high in sulfur and potassium, a mild stimulant.

NUTRIENTS IN ONE POUND

Calories	84	Iron	9.1 mg
Protein	7.7 g	Vitamin A	20,450 I.U.
Fat	1.4 g	Thiamine	0.37 mg
Carbohydrates	15.0 g	Riboflavin	0.71 mg
Calcium	885 mg	Niacin	3.6 mg
Phosphorus	209 mg	Ascorbic acid	350 mg

Appendix A

Food Analysis Chart

	Acid	Alkaline	Proteins	Starches	Fats	Vegetables	Fruits	Hours of Time Required for Digestion	Caloric Value Per Pound	Vitamins found in these Foods
Agar-agar		X				X		1½	75	E
Almonds		X			X			2½	3000	A B
Apples		X					X	2¾	275	A B C
Apricots		X					X	2¾	300	B C
Artichokes		X				X		2	350	A B C
Asparagus	X					X		2¼	150	A B C
Avocados—										
Alligator pears		X			X	X		1¾	835	A B C
Bananas		X	X				X	3	350	A B C E
Barley—whole grain	X			X				3¾	1500	A B E
Barley—pearled	X			X				4	1250	B
Beans—dried		X		X		X		3	750	A B E
Beans—fresh string		X				X		3¼	100	A B C
Beans—green Lima		X				X		2½	600	A B C
Beans—dried Lima		X		X		X		2½	750	B E
Beans—kidney		X		X		X		3	750	B E
Beans—soy		X		X		X		3	750	B E
Beechnuts	X		X		X			3	2800	A B
Beef—lean	X		X					3½	1000	E
Beef—dried	X		X					3½	1000	E
Beef—raw juice	X		X					3	1000	E
Beets		X		X		X		2¾	150	A B C
Beet tops		X				X		2	150	A B C E
Blackberries	X						X	2½	250	B C
Blueberries	X						X	2	300	B C
Bran		X						2¾	1100	B E
Broccoli		X				X		3	160	A B C
Brussel sprouts		X		X		X		4	250	A B C
Butter	X				X			3¼	3700	A D E
Buttermilk		X			X			2¼	150	A B C D
Cabbage—white, raw		X				X		3	160	A B C E
Cabbage—red, raw		X				X		3¾	160	A B C E
Cantaloupe		X					X	3¼	300	A B C
Carrots		X		X		X		2¼	700	A B C D
Carrots—raw		X				X		3	240	A B C D

	Acid	Alkaline	Proteins	Starches	Fats	Vegetables	Fruits	Hours of Time Required for Digestion	Caloric Value Per Pound	Vitamins found In these Foods
Cashew nuts	X				X			3¼	2500	A B
Cauliflower		X		X		X		2¼	150	A B C
Celery		X				X		3¼	90	A B C
Celery root		X		X		X		3½	250	A B C
Cereals—whole grain	X			X				3	1800	B D E
Cheese	X				X			3¼	2000	A B
Cherries—red		X					X	2	350	B C
Cherries—white		X					X	2	350	B C
Chestnuts	X				X			2¾	2500	A B
Chicken	X		X					3¼	500	E
Chocolate	X			X				2	2900	B
Cider	X						X	1¼	275	B C
Clams—round	X		X					4	200	A B D
Clams—long soft	X		X					3¾	200	A B D
Cocoa	X			X				2	2400	B
Cocoanut—dried natural		X			X		X	3¼	2800	A B
Cocoanut—fresh		X			X		X	2¾	2000	A B
Cocoanut milk		X			X		X	2	500	A B
Cod liver oil	X		X		X			3	2500	A D E
Cottage Cheese	X				X			3¼	2100	A B C D
Corn—sweet		X		X		X		3	490	A B C
Corn—dried		X		X		X		2¾	800	A B C
Cornmeal	X			X				3½	1800	A B C
Cottonseed meal	X			X				3¾	1000	A D E
Cow peas	X			X		X		3¼	600	A D E
Cranberries		X					X	3¼	225	B C
Cream	X				X			2½	900	A B D
Cucumbers		X				X		3¼	100	A B C
Currants—fresh	X						X	3	300	B C
Currants—dried		X					X	2½	1500	B C
Dandelion		X				X		2½	235	A B C E
Dates		X		X			X	2½	1600	A B C
Dill		X				X		3¼	250	A B C
Egg plant		X		X		X		3½	150	A B C
Egg whites	X							2¼	300	
Egg yolks	X				X			2¼	1700	A B D E
Endive		X				X		3	100	A B C
Farina	X			X				3½	900	B E
Figs—dried		X					X	2½	1400	B C
Figs—fresh		X					X	2¼	350	B C
Flaxseed		X				X		3	875	B E
Flour—buckwheat	X			X				4	1000	B E
Flour—gluten	X				X			3¼	600	B E
Flour—graham	X			X				3	1800	B
Flour—rye	X			X				3¼	1100	A B
Flour—white	X			X				4	12000	
Flour—whole wheat	X			X				3	1700	B D E
Frog's legs	X		X					4	500	E
Garlic		X				X		2	200	A B C
Gooseberries	X						X	2½	300	B C
Gluten feed	X				X			3	600	B E
Grapes—white	X						X	1¾	350	A B C
Grapes—Concord	X						X	1¾	350	A B C
Grapefruit		X					X	2	200	A B C
Guava		X			X		X	3	450	B C
Halibut	X		X					2½	450	E
Ham	X		X		X			4	1900	A B
Hazelnuts	X				X			3	2400	A B
Herbs		X				X		2½	200	A B C
Hominy	X			X				3	800	B E
Honey	X			X				2¼	1600	B
Horseradish	X					X		4	100	A B C

	Acid	Alkaline	Proteins	Starches	Fats	Vegetables	Fruits	Hours of Time Required for Digestion	Caloric Value Per Pound	Vitamins found in these Foods
Huckleberries	X						X	2¾	200	B C
Irish moss		X				X		1½	100	E
Kohlrabi	X			X		X		3	200	A B C
Kelp salt		X				X		3	100	E
Lamb	X		X		X			3	1000	A B
Leeks		X				X		2½	200	A B C E
Lemons		X					X	1½	200	B C
Lentils		X		X		X		3	1700	A B
Lettuce		X				X		2¼	100	A B C E
Limes		X					X	3	200	A B C
Linseed meal	X			X		X		4	1000	B E
Lupins—dried		X		X		X		2¾	1500	B E
Mackerel	X		X					3¼	450	E
Macaroni—white	X			X				3¾	1200	B
Macaroni—whole wheat	X			X				3	1700	B E
Mango		X			X		X	1¾	450	B C
Maple syrup	X			X				1¼	1400	B
Milk—cow's (skimmed)		X			X			2	300	A B C
Milk—(cow's whole raw)		X			X			2¼	1200	A B CDE
Milk—condensed	X			X				4	1500	A B C
Milk—human		X			X			1½	1800	A B CDE
Milk—goat's		X			X			2	1600	A B CDE
Millet	X			X				3¼	950	A
Molasses	X			X				2¼	1800	B
Mushrooms		X		X		X		2½	200	A B
Muskmelons		X					X	3¼	300	A B C
Mustard greens		X				X		3½	150	A B C E
Oats—whole grain	X			X				3¼	1900	A B E
Oatmeal	X			X				3½	1800	B E
Okra		X		X		X		2½	200	A B C
Olives—ripe		X			X		X	1¾	400	A E
Olives—dried		X			X		X	2	800	A E
Olive oil		X			X			3¼	4000	A D E
Onions		X				X		3¼	200	A B C
Oranges		X					X	2	200	A B CDE
Parsley		X				X		1½	270	B
Parsnips		X		X		X		3½	380	A B
Peaches—fresh	X						X	2½	200	A B C
Peaches—dried		X					X	2¾	1200	B C
Peanuts	X				X			3¼	3000	A B E
Pears	X						X	2¼	350	B
Peas—fresh green		X		X		X		3¼	500	A B C
Peas—dried		X		X		X		3½	1500	B E
Pecan nuts	X				X			2¾	3500	A B
Peppers—fresh green		X				X		3¼	450	A B C
Persimmons		X					X	3¾	350	B C
Pignolias	X				X			2¾	2700	A B
Pineapple		X					X	2¼	200	A B C
Plums	X						X	2¾	250	B C
Pomegranates	X						X	3¼	450	B C
Pork	X		X		X			4	1200	A B
Potatoes—Irish		X		X		X		2	400	A B C
Potatoes—sweet		X		X		X		3¼	400	A B C
Prunes—fresh	X						X	2¾	350	A B
Prunes—dried	X						X	3	1550	A B
Pumpkin		X		X		X		3¼	160	A B C
Peppermint leaves		X				X		2½	100	A B C E
Quinces	X						X	3¾	400	B C
Radishes		X				X		3¼	100	B C

	Acid	Alkaline	Proteins	Starches	Fats	Vegetables	Fruits	Hours of Time Required for Digestion	Caloric Value Per Pound	Vitamins found in these Foods
Raisins		X		X			X	2	1650	B C
Raspberries	X						X	1¾	150	B C
Rhubarb	X						X	3	1000	B C
Rice—natural brown	X			X				2	1600	A B E
Rice—white	X			X				2½	800	
Romaine		X				X		2¼	100	A B C E
Roquefort cheese	X				X			3¾	1600	B
Rutabagas	X			X		X		3¼	200	A B C
Rye—whole	X			X				3½	1100	B E
Sauerkraut	X					X		4½	200	A B C
Sage		X				X		2¾	150	A B C E
Sago	X			X				3	1800	B
Salmon	X		X					3¾	450	A B D
Salmon—smoked	X		X					4½	400	E
Scallops	X		X					3¼	450	E
Sea grass		X				X		1½	100	E
Smelt	X		X					3½	450	E
Smoked herring	X		X					4½	450	A B D
Sorrel		X				X		2¾	100	A B C E
Sole	X		X					2¾	450	E
Spaghetti—white	X			X				3½	1200	B
Spaghetti—whole wheat	X			X				3	1700	B E
Spinach		X				X		3	100	A B C D
Squab	X		X					3¼	1000	E
Squash—Italian		X				X		3	150	A B C
Squash—summer		X				X		2¾	150	A B C
Squash—yellow crook neck		X				X		3	150	A B C
Squash—Hubbard		X		X		X		2¼	200	A
Squash—banana		X		X		X		2¾	200	A
Squash—other winter varieties		X		X		X		3	200	A
Strawberries	X						X	2¼	150	A B C
Sugar—raw		X		X				1¼	1800	B
Sugar—white	X			X				1¼	1500	B
Swiss chard		X				X		3	150	A B C D
Syrup	X			X				1¼	1400	A B
Tapioca	X			X				2½	1800	B
Tomatoes—cooked	X					X		1¾	100	A B C
Tomatoes—raw		V				X		2	100	A B C
Turnips		X		X		X		4	200	A B C
Turnip tops		X				X		3¼	150	A B C E
Turkey	X		X					3¼	1000	E
Vegex		X				X		2½	1000	A B E
Vegetable oils		X			X			3½	2000	A D E
Vinegar—cider	X							2¼	100	
Walnuts	X				X			3	3000	B
Watercress		X				X		3¼	100	A B C
Watermelon		X					X	2¾	120	A B C
Whole wheat grain	X			X				3¾	1700	B D E
Wheat bran		X		X				2¾	1100	B
Wheat germ	X			X				3	500	A B E
Wheat gluten	X				X			3¼	600	B E
Whey—cow's milk		X		X				3	150	A B C
Whey—goat's milk		X		X				3	150	A B C
Wine	X						X	1¼	150	C
Wild berries	X						X	2	150	B C
Whiting fish	X		X					3¼	50	E
Wintergreen		X				X		2¾	100	A B C E

Index

About the Author

Bernard Jensen was born March 25, 1908, in Stockton, California. Following in his father's footsteps, he entered the West Coast Chiropractic College in Oakland, California at the age of 18. Shortly after his graduation, his health failed. Doctors diagnosed the problem as bronchiectasis, a severe and often fatal lung condition for which there was no known cure. He was told that medical science could offer him no hope.

Unwilling to give up, young Dr. Jensen sought the help of a Seventh-Day Adventist physician who taught him the basics of proper nutrition, pulled him off all "junk" foods, and placed him on a natural food diet. Improvement was slow, but steady. As a result of a breathing exercise program developed by Thomas Gaines, he added four inches to his chest in a year's time. Because nature offered him a cure when medical science could do nothing, Dr. Jensen determined to learn all he could about natural healing.

His training included postgraduate courses at the National Chiropractic College in Chicago, Illinois. Later, Dr. Jensen returned to California to study iridology, the science of interpreting tissue conditions from the iris, with Dr. R.M. McLain at the International School of Arts and Sciences in San Francisco, California.

To expand his knowledge of health work, Dr. Jensen studied bowel care with Dr. John Harvey Kellogg of Battle Creek, Michigan, and Dr. Max Gerson of New York, the latter known for his use of nutrition, diets, supplements, and enemas in the treatment of degenerative disease. Among others, Dr. Jensen also studied with Dr. O.B. Shellberg of New York, a colonics specialist; Dr. Ralph Benner, of the Bircher-Benner Clinic in Switzerland; Dr. John Tilden of Denver, Colorado, and Dr. George Weger of Redlands, California.

Dr. Jensen operated several health sanitariums in California, the first in Ben Lomond, the second at Alta Dena, and the last at Escondido. It is this last sanitarium that he refers to as "the Ranch." At the sanitariums, he lived with his patients day in and day out. "The sanitarium was my university," Dr. Jensen says, "and patients were my books." The sanitariums were living laboratories where he was able to observe firsthand what best brought patients back to health.

Over the years, Dr. Jensen has received many honors and awards, including Knighthood in the Order of St. John of Malta; the Dag Hammarskjöld Peace Award of the Pax Mundi Academy in Brussels, Belgium; and an award from Queen Juliana of the Netherlands for his nutritional work. In 1982, he also received the National Health Federation's Pioneer Doctor of the Year award.

At the age of 76, Dr. Jensen earned his Ph.D. from the University of Humanistic Studies in San Diego, California, climaxing a lifetime of study, work, and teaching in the healing arts. He has lectured in 55 countries around the world and has authored over 40 books on the subjects of natural health care and iridology.

Now in his eighties, Dr. Jensen continues to write, lecture, travel, and learn.

If You've Enjoyed Reading This Book . . .

. . . why not tell a friend about it? If you're interested in learning more about Dr. Bernard Jensen's approach to health, here are some other titles you may find to be informative, engaging, and fun.

Vibrant Health From Your Kitchen

A warm and wonderful tour through Dr. Jensen's latest discoveries about food, nutrition, and health, this book provides the guidance needed to keep your family disease-free, healthy, and happy.

Tissue Cleansing Through Bowel Management

Toxin-laden tissue can become a breeding ground for disease. This remarkable book instructs you in the removal of toxins and the restoration of health and youthfulness through the cleansing and care of the organs of elimination.

Food Healing for Man

We now know that foods can repair the tissue damage that accompanies most illness and disease. Look over the shoulders of the great pioneer nutritionists as they investigate the links between nutrition and disease.

Chlorella: Gem of the Orient

Why does Dr. Jensen consider chlorella—a green alga—the most valuable broad-spectrum food supplement discovery of the twentieth century? You'll find out in this unusually beautiful, fully illustrated, hard cover book.

Creating a Magic Kitchen

This is Dr. Jensen's introductory primer on the art of selecting and preparing foods for the best of health. Short, easy to understand, and handy to use, this is the perfect book for anyone who wants a more healthful and enjoyable lifestyle.

Nature Has a Remedy

This popular classic provides a delightful description of the many paths to natural healing—foods, herbs, exercise, climate selection, personology, and hundreds of effective remedies.

World Keys to Health and Long Life

Based on Dr. Jensen's travels to over fifty-five countries, this fascinating book describes the health and longevity secrets of centenarians interviewed in the Hunza Valley of India; Vilcabamba, Peru; the Caucasus Mountains of the Soviet Union; and other places around the world.

Doctor-Patient Handbook

Discover the reversal process and healing crisis that Nature uses to rid the body of disease and restore well-being. Here is a fresh approach to wholistic health.

Slender Me Naturally

Dr. Jensen's answer to fad diets that don't work is a natural weight loss program that does. Developed over fifty-eight years of experience with overweight patients, this program is a healthful and effective way of losing unwanted weight.

Breathe Again Naturally

Get rid of asthma, allergies, bronchitis, hay fever, and other respiratory problems. Dr. Jensen discusses nutrition, herbs that work, food supplements, breathing exercises, attitude, and climate.

Arthritis, Rheumatism and Osteoporosis

Are you among the one in four Americans who suffers from arthritis, rheumatism, or osteoporosis? Would you like to know what to do about it? This book is for you.

In Search of Shangri-La

Here is the very personal journal of Dr. Jensen's physical and spiritual travels through China into Tibet, and his reflections on his search for Shangri-La.

Beyond Basic Health

Dr. Jensen looks at the deteriorating state of modern man's health and offers practical advice and insights to those health professionals who must deal with today's devastating illnesses.

Love, Sex and Nutrition

Based on years of detailed study, this book explores the link between diet, sensuality, and relationships. This is an important and practical guide for people who wish to improve their sexuality safely and naturally.

For information regarding prices, write to:

> Hidden Valley Health Ranch
> Route 1 Box 52
> Escondido, California 92025